# Advanced Theoretical and Computational Methods for Complex Materials and Structures

# Advanced Theoretical and Computational Methods for Complex Materials and Structures

Editors

**Francesco Tornabene**
**Rossana Dimitri**

MDPI • Basel • Beijing • Wuhan • Barcelona • Belgrade • Manchester • Tokyo • Cluj • Tianjin

*Editors*
Francesco Tornabene
Department of Innovation Engineering,
University of Salento
Italy

Rossana Dimitri
Department of Innovation Engineering,
University of Salento
Italy

*Editorial Office*
MDPI
St. Alban-Anlage 66
4052 Basel, Switzerland

This is a reprint of articles from the Special Issue published online in the open access journal *Applied Sciences* (ISSN 2076-3417) (available at: https://www.mdpi.com/journal/applsci/special_issues/complex_materials).

For citation purposes, cite each article independently as indicated on the article page online and as indicated below:

| |
|---|
| LastName, A.A.; LastName, B.B.; LastName, C.C. Article Title. *Journal Name* **Year**, *Volume Number*, Page Range. |

**ISBN 978-3-0365-1118-4 (Hbk)**
**ISBN 978-3-0365-1119-1 (PDF)**

# Contents

# About the Editors

**Francesco Tornabene** is a researcher at the School of Engineering, Department of Innovation Engineering, University of Salento. He was born on 13 January 1978 in Bologna, where he received the high school degree at Liceo Classico San Luigi, in 1997. In 2001 he achieved a National Patent Bologna (Italy) for the Industrial Invention: Friction Clutch for High Performance Vehicles Question BO2001A00442. He received from the University of Bologna—Alma Mater Studiorum, a M.Sc. degree in Mechanical Engineering (Curriculum in Structural Mechanics) on 23/07/2003, discussing a thesis entitled: "Dynamic Behavior of Cylindrical Shells: Formulation and Solution". In December 2003, he was admitted at the PhD course in Structural Mechanics, at the University of Bologna; in 2004, he received from the University of Bologna a Thesis prize in memory of Carlo Felice Jodi; in 2007, he received the Ph.D. degree in Structural Mechanics at the University of Bologna, discussing the Thesis entitled "Modeling and Solution of Shell Structures Made of Anisotropic Materials". From 2007 to 2009 he received a research fellowship by the University of Bologna, working on the "Unified Formulation of Shell Structures Made of Anisotropic Materials. Numerical Analysis Using the Generalized Differential Quadrature Method and the Finite Element Method". From 2011 to 2012, he became a junior researcher within the research program entitled "Advanced Numerical Schemes for Anisotropic Materials"; from 2012 to 2018, he was an Assistant Professor and Lecturer at the Alma Mater Studiorum—University of Bologna; from 2018 to date he is an Aggregate Professor in Structural Mechanics and Lecturer at the University of Salento, Department of Innovation Engineering (Lecce). For a long time, his scientific interests include structural mechanics, solid mechanics, innovative and smart materials, computational mechanics and numerical techniques, and damage and fracture mechanics. He is author of more than 280 scientific publications, and collaborates with many national or international researchers and professors all over the world, as visible from his scientific production. He is author of 11 books; see, for example, "Meccanica delle Strutture a Guscio in Materiale Composito. Il metodo Generalizzato di Quadratura Differenziale" (2012); "Mechanics of Laminated Composite Doubly-Curved Shell Structures. The Generalized Differential Quadrature Method and the Strong Formulation Finite Element Method" (2014); "Laminated Composite Doubly-Curved Shell Structures I. Differential Geometry. Higher-Order Structural Theories" (2016); "Laminated Composite Doubly-Curved Shell Structures II. Differential and Integral Quadrature. Strong Formulation Finite Element Method" (2016); "Anisotropic Doubly-Curved Shells. Higher-Order Strong and Weak Formulations for Arbitrarily Shaped Shell Structures" (2018); among many others. He is member of the Editorial Board for 44 international journals. He is also Editor-in-Chief for two international journals: *Curved and Layered Structures and Journal of Composites Science*; furthermore, he is Associate Editor for seven international journals. In the last few years, he has received different important awards, for example, "Highly Cited Researcher by Clarivate Analytics" (years 2018, 2019, 2020), "Ambassador of Bologna Award for the organization of 21st International Conference on Composite Structures ICCS21, 4-7 September 2018, Bologna, Italy" (2019), and "Member of the European Academy of Sciences" (since 2018). He collaborates as reviewer with more than 260 prestigious international journals in the structural mechanics field. From 2012, his teaching activity has included dynamics of structures; computational mechanics; plates and shells; theory of structures; structural mechanics. He is habilitated as Associate Professor and Full Professor in the area 08/B2 (Structural Mechanics) and as Associate Professor in Area 09/A1 (aeronautical and aerospace engineering and naval architecture).

**Rossana Dimitri** is an Associate Professor at the School of Engineering, Department of Innovation Engineering, University of Salento, Lecce, Italy (since 2019). She received from the University of Salento, a M. Sc. degree in "Materials Engineering" in 2004, a Ph.D. degree in "Materials and Structural Engineering" in 2009, and a Ph.D. degree in "Industrial and Mechanical Engineering" in 2013. In 2005, she received from the University of Salento the "Best M. Sc. Thesis Prize 2003–2004" in memory of Eng. Gabriele De Angelis; in 2013, she was awarded by the Italian Group for Computational Mechanics (GIMC) for the Italian selection of the 2013 ECCOMAS PhD Award. Her current interests include structural mechanics, solid mechanics, damage and fracture mechanics, contact mechanics, isogeometric analysis, high-performing computational methods, and consulting in applied technologies and technology transfer. During 2010 and 2011, she received a research fellowship by ENEA Research Centre of Brindisi (UTTMATB-COMP) for the development and the characterization of some thermoplastic composites for thermal solar panels and adhesively bonded turbine blades under severe environmental conditions. During 2011 and 2012, she was a visiting scientist with a fellowship at the Institut für Kontinuumsmechanik Gottfried Wilhelm Leibniz Universität Hannover to study interfacial problems with isogeometric approaches. From 2013 to 2016, she was a researcher at the University of Salento, within the ERC starting research grant "INTERFACES" on "Computational mechanical modelling of structural interfaces based on isogeometric approaches". From 2013 to 2019, she has collaborated as a researcher RTD-B with the University of Bologna and the Texas A&M University for a comparative assessment of some advanced numerical collocation methods with lower computational cost for fracturing problems and structural modelling of composite plates and shells, made by isotropic, orthotropic and anisotropic materials. She is author of 112 scientific publications, and she collaborates with many national or international researchers and professors worldwide, as visible from her scientific production. She also collaborates with different prestigious international journals in the structural mechanics field, as reviewer, member of the editorial board, and guest editor for different Special Issues.

*applied sciences*

MDPI

*Editorial*

# Special Issue on Advanced Theoretical and Computational Methods for Complex Materials and Structures

**Francesco Tornabene * and Rossana Dimitri ***

Department of Innovation Engineering, Università del Salento, Via per Monteroni, 73100 Lecce, Italy
* Correspondence: francesco.tornabene@unisalento.it (F.T.); rossana.dimitri@unisalento.it (R.D.)

## 1. Introduction

The large use of composite materials and shell structural members with complex geometries in technologies related to various branches of engineering, has gained increased attention from scientists and engineers for the development of even more refined approaches, to investigate their mechanical behavior. It is well known that composite materials are able to provide higher values of strength stiffness, and thermal properties, together with conferring reduced weight, which can affect the mechanical behavior of beams, plates, and shells, in terms of static response, vibrations, and buckling loads. At the same time, enhanced structures made of composite materials can feature internal length scales and non-local behaviors, with great sensitivity to different staking sequences, ply orientations, agglomeration of nanoparticles, volume fractions of constituents, and porosity levels, among others. In addition to fiber-reinforced composites and laminates, increased attention has been paid in literature to the study of innovative components such as functionally graded materials (FGMs), carbon nanotubes (CNTs), graphene nanoplatelets, and smart constituents. Some examples of smart applications involve large stroke smart actuators, piezoelectric sensors, shape memory alloys, magnetostrictive and electrostrictive materials, as well as auxetic components and angle-tow laminates. These constituents can be included in the lamination schemes of smart structures to control and monitor the vibrational behavior or the static deflection of several composites. The development of advanced theoretical and computational models for composite materials and structures is a subject of active research and this is explored here for different complex systems, including their static, dynamic, and buckling responses; fracture mechanics at different scales; as well as the adhesion, cohesion, and delamination of materials and interfaces.



**Citation:** Tornabene, F.; Dimitri, R. Special Issue on Advanced Theoretical and Computational Methods for Complex Materials and Structures. *Appl. Sci.* **2021**, *11*, 2532. https://doi.org/10.3390/app11062532

Received: 6 March 2021
Accepted: 10 March 2021
Published: 12 March 2021



## 2. Enhanced Theoretical and Computational Formulations

In a context where an increased theoretical/computational demand is usually required to solve solid mechanics problems, this Special Issue has collected 11 papers on the application of high-performing computational strategies and enhanced theoretical formulations to solve different static and/or dynamic problems, for different engineering applications also in coupled conditions. A wide variety of examples and topics is considered, with highly useful insights both from a scientific and design perspective. More specifically, the first paper, authored by M.H. Jalaei, R. Dimitri and F. Tornabene, combines the Hamilton's variational principle and the Eringen's constitutive model to study the dynamic stability of orthotropic temperature-dependent single-layered graphene sheets embedded in a temperature-dependent elastomeric medium and subjected to a biaxial oscillating loading in thermal environment [1]. These nanostructures are largely adopted as important components in various highly technological industries, such as nanoactuators, nanoresonators, nanosensors, and nanocomposites, such that non-classical continuum approaches are usually required for an accurate analysis of results, especially for complicated coupled problems.

In the second work by K.J. Huang et al. [2], the authors develop an efficient theoretical approach to investigate the nonlocal and size-dependent (NLSD) effects on the dielectric response of plasmonic nanostructures that incorporate the spatially local, nonlocal, size and even analogous quantum-size responses of the material. On the other hand, unlike traditional engineering structural problems, the design of micro-electromechanical systems (MEMS) usually involves microstructures, novel materials, and extreme operating conditions, where multi-source uncertainties usually exist. In such a context, the work by Z. Huang et al. [3] develops an evidence–theory-based robustness optimization method for a robust design of MEMS, including a micro-force sensor, an image sensor, and a capacitive accelerometer of practical interest. Another application is represented by microfluidic devices, for which an analytical and practical method is efficiently proposed in the work by U. Tuzun [4] to study the Brownian motion in nanoparticle suspensions, as used in a variety of applications involving the transport of reagents and products to and from structured material surfaces. Among geotechnical engineering applications, the work authored by H. Sun and W. Sun [5] provides a useful finite element study for an effective measurement of safety and serviceability of existing metro tunnels during adjacent excavation, in lieu of centrifuge model tests, analytical or semi-analytical methods, or a more expensive in-situ monitoring. In the further work by Y. Zhu et al. [6], a quantitative evaluation of the solid displacement induced by a ground loss and shield machine mechanical effect is provided in metro tunnel constructions, based on an elastic half-space Mindlin model. The proportion of stratum displacement caused by each factor was quantitatively analyzed in this work, as a valid guidance for the stratum displacement calculation of shield tunneling in the future. A novel method is also proposed by X. Tan, H. Zhang et al. [7] to define the three-dimensional Greenfield stratum movements caused by a shield tunnel construction, where an elastic half-space model of mirror-sink method is combined with a modified analytical approach to give a useful design tool for underground engineering constructions. An efficient numerical model is, instead, established by H. Sun et al. [8] to compute the tunnel deformation caused by a circular excavation, while adopting a hypoplasticity nonlinear constitutive model, able to account for path-and strain-dependent soil stiffnesses even at small strains. This method is useful for practical engineering soil-structure problems. The extensive use of composite materials and structures in many engineering applications with complex microstructure and manufacturing processes, requires a thorough attention to their mechanical performances, such as the structural deflection damage and load capacity [9–12], as well as the buckling and dynamic behavior [13–18], along with possible related uncertainties and stochastic variations. In this setting, the work authored by S. Zhang and X. Chen [19] provides a stochastic natural frequency analysis of typical composite structures with micro-scale (constituent-scale) and meso-scale (ply-scale) uncertainties, based on Monte-Carlo simulations and the response surface method. Another numerical model is proposed by S. Shahbazi et al. [20] to analyze the non-linear time-history of different structural members under the simultaneous effects of far- and near-field earthquakes, as useful for design computational tools. The last work, authored by A. Pavlovic, et al. [21] finally proposes a finite element study of the mechanical behavior of palletized products and polyethylene terephthalate bottles, partially filled with liquid under compressive stress conditions, as commonly occurs during a transportation process. In this last case, an accurate computational prediction of the transport-related integrity risks of bottles is desirable in order to avoid any kind of instability and damage phenomena within commercial products, with a clear economical repercussion.

## 3. Future Developments

Although this Special Issue has been closed, further developments on the theoretical and computational modeling of enhanced structures and composite materials are expected, including their static, dynamic, and buckling responses; fracture mechanics at different scales; as well as the adhesion, cohesion, and delamination of materials and interfaces, as useful for many industrial applications.

## References


1. Jalaei, M.H.; Dimitri, R.; Tornabene, F. Dynamic Stability of Temperature-Dependent Graphene Sheet Embedded in an Elastomeric Medium. *Appl. Sci.* **2019**, *9*, 887. [CrossRef]
2. Huang, K.J.; Qin, S.J.; Zhang, Z.P.; Ding, Z.; Bai, Z.C. Nonlocal and Size-Dependent Dielectric Function for Plasmonic Nanoparticles. *Appl. Sci.* **2019**, *9*, 3083. [CrossRef]
3. Huang, Z.; Xu, J.; Yang, T.; Li, F.; Deng, S. Evidence-Theory-Based Robust Optimization and Its Application in Micro-Electromechanical Systems. *Appl. Sci.* **2019**, *9*, 1457. [CrossRef]
4. Tuzun, U. Improving Conductivity in Nano-Conduit Flows by Using Thermal Pulse-Induced Brownian Motion: A Spectral Impulse Intensity Approach. *Appl. Sci.* **2019**, *9*, 3889. [CrossRef]
5. Sun, H.; Sun, W. Effect of Soil Reinforcement on Tunnel Deformation as a Result of Stress Relief. *Appl. Sci.* **2019**, *9*, 1420. [CrossRef]
6. Zhu, Y.; Chen, L.; Zhang, H.; Tu, P.; Chen, S. Quantitative Analysis of Soil Displacement Induced by Ground Loss and Shield Machine Mechanical Effect in Metro Tunnel Construction. *Appl. Sci.* **2019**, *9*, 3028. [CrossRef]
7. Tan, X.; Zhang, H.; Zhang, G.; Zhu, Y.; Tu, P. An Improved Method for Predicting the Greenfield Stratum Movements Caused by Shield Tunnel Construction. *Appl. Sci.* **2019**, *9*, 4522. [CrossRef]
8. Sun, H.; Wang, L.; Chen, S.; Deng, H.; Zhang, J. A Precise Prediction of Tunnel Deformation Caused by Circular Foundation Pit Excavation. *Appl. Sci.* **2019**, *9*, 2275. [CrossRef]
9. Tornabene, F.; Viola, E. Static analysis of functionally graded doubly-curved shells and panels of revolution. *Meccanica* **2013**, *48*, 901–930. [CrossRef]
10. Dimitri, R.; Tornabene, F.; Zavarise, G. Analytical and numerical modeling of the mixed-mode delamination process for composite moment-loaded double cantilever beams. *Compos. Struct.* **2018**, *187*, 535–553. [CrossRef]
11. Tornabene, F. On the critical speed evaluation of arbitrarily oriented rotating doubly-curved shells made of functionally graded materials. *Thin Walled Struct.* **2019**, *140*, 85–98. [CrossRef]
12. Dimitri, R.; Tornabene, F.; Reddy, J.N. Numerical study of the mixed-mode behavior of generally-shaped composite interfaces. *Compos. Struct.* **2020**, *237*, 111935. [CrossRef]
13. Nejati, M.; Dimitri, R.; Tornabene, F.; Yas, M.H. Thermal buckling of nanocomposite stiffened cylindrical shells reinforced by functionally Graded wavy Carbon NanoTubes with temperature-dependent properties. *Appl. Sci.* **2017**, *7*, 1223. [CrossRef]
14. Noroozi, A.R.; Malekzadeh, P.; Dimitri, R.; Tornabene, F. Meshfree radial point interpolation method for the vibration and buckling analysis of FG-GPLRC perforated plates under an in-plane loading. *Eng. Struct.* **2020**, *221*, 111000. [CrossRef]
15. Sofiyev, A.H.; Tornabene, F.; Dimitri, R.; Kuruoglu, N. Buckling behavior of FG-CNT reinforced composite conical shells subjected to a combined loading. *Nanomaterials* **2020**, *10*, 419. [CrossRef] [PubMed]
16. Tornabene, F.; Ceruti, A. Mixed static and dynamic optimization of four-parameter functionally graded completely doubly curved and degenerate shells and panels using GDQ method. *Math. Prob. Eng.* **2013**, 867079. [CrossRef]
17. Tornabene, F.; Dimitri, R.; Viola, E. Transient dynamic response of generally-shaped arches based on a GDQ-time-stepping method. *Int. J. Mech. Sci.* **2016**, *114*, 277–314. [CrossRef]
18. Malikan, M.; Dimitri, R.; Tornabene, F. Effect of sinusoidal corrugated geometries on the vibrational response of viscoelastic nanoplates. *Appl. Sci.* **2018**, *8*, 1432. [CrossRef]
19. Zhang, S.; Chen, X. Stochastic Natural Frequency Analysis of Composite Structures Based on Micro-Scale and Meso-Scale Uncertainty. *Appl. Sci.* **2019**, *9*, 2603. [CrossRef]
20. Shahbazi, S.; Karami, A.; Hu, J.W.; Mansouri, I. Seismic Response of Steel Moment Frames (SMFs) Considering Simultaneous Excitations of Vertical and Horizontal Components, Including Fling-Step Ground Motions. *Appl. Sci.* **2019**, *9*, 2079. [CrossRef]
21. Pavlovic, A.; Fragassa, C.; Vegliò, L.; Vannucchi de Camargo, F.; Minak, G. Modeling Palletized Products: The Case of Semi-Filled Bottles under Top-Load Conditions, Including Fling-Step Ground Motions. *Appl. Sci.* **2020**, *10*, 332. [CrossRef]

*applied sciences*

MDPI

*Article*

# Dynamic Stability of Temperature-Dependent Graphene Sheet Embedded in an Elastomeric Medium

**Mohammad Hossein Jalaei [1], Rossana Dimitri [2] and Francesco Tornabene [2,*]**

1 Young Researchers and Elite Club, Islamshahr Branch, Islamic Azad University, Islamshahr 67653-33147, Iran; m.h.jalaei@gmail.com

2 Department of Innovation Engineering, Università del Salento, 73100 Lecce, Italy; rossana.dimitri@unisalento.it

* Correspondence: francesco.tornabene@unisalento.it

Received: 30 January 2019; Accepted: 25 February 2019; Published: 1 March 2019

**Abstract:** This work applies the first-order shear deformation theory (FSDT) to study the dynamic stability of orthotropic temperature-dependent single-layered graphene sheet (SLGS) embedded in a temperature-dependent elastomeric medium and subjected to a biaxial oscillating loading in a thermal environment. Possible thermal effects are considered in the size-dependent governing equations of the problem. These last ones are derived by means of the Hamilton's variational principle combined with the Eringen's differential constitutive model. Navier's solution as well as Bolotin's approach are applied to obtain the dynamic instability region (DIR) of the graphene sheet. Thus, a parametric study is carried out to explore the sensitivity of the DIR of the graphene sheet to the temperature variation, the static load factor, the aspect ratio, the foundation type, and the nonlocal parameter (NP). Results indicate that the dimensionless pulsation frequency reduces for increasing values of temperature and NP, whereas the size effect becomes even more pronounced for increasing temperatures. In addition, the adoption of temperature-dependent mechanical properties, rather than independent ones, yields a global shift of the DIR to smaller pulsating frequencies. This proves the relevance of the temperature-dependent mechanical properties to obtain reliable results, in a physical sense.

**Keywords:** dynamic stability; elastomeric foundation; Eringen's differential constitutive model; graphene sheet; temperature-dependent properties

## 1. Introduction

Among nanostructures, graphene sheets have increased the interest of the scientific community, owing to their astonishing thermal, chemical, electrical, and mechanical properties. Due to the distinguishing features, theses nanostructures have been adopted as important components in various high technology industries such as nanoactuators, nanoresonators, nanosensors, and nanocomposites.

In order to investigate the mechanical characteristics of micro/nano-scale structures accurately, the small-scale effects should be considered. It has been recognized that the main properties of materials and structures at micro/nano-scale are size-dependent and differ significantly from their behavior at larger scales. Hence, non-classical continuum mechanics such as couple stress theory (CST), nonlocal strain gradient theory (NSGT), and nonlocal elasticity theory (NET) have been applied by the scientific community to analyze the mechanical behavior of micro/nanomaterials due to the difficulties in experimental investigations at the nanoscale and the high computational costs of molecular dynamics (MD) simulations.

Among these theories, the NET introduced by Eringen [1] has been broadly employed in order to include the size effects on the static and dynamic analyses of nanomaterials. For example, Murmu et al. [2] investigated the in-plane magnetic field effect on the vibration of a magnetically sensitive single-layered graphene sheet (SLGS) resting on an elastic foundation via the NET in conjunction with Kirchhoff plate theory. Wang et al. [3] presented an analytical solution for the nonlinear vibration behavior of viscoelastic double layered nanoplates. The problem was tackled through the Kirchhoff plate theory and the multiple scales method. Karličić et al. [4] performed the nonlinear vibration and dynamic instability of single-walled carbon nanotubes (SWCNTs) resting on a Kelvin–Voigt viscoelastic medium under magnetic field using NET and the Euler–Bernoulli beam theory (EBT). Arani and Jalaei [5] employed the Fourier–Laplace transformation method as well as the sinusoidal shear deformation theory (SSDT) to examine the longitudinal magnetic field effect on the transient behavior of a viscoelastic simply-supported orthotropic nanoplate embedded on a visco-Pasternak medium. Kolahchi et al. [6] studied the dynamic buckling of embedded viscoelastic laminated nanoplates on the basis of NET in conjunction with refined zigzag theory. Jalaei et al. [7] studied the dynamic stability of a viscoelastic graphene sheet embedded on viscose matrix with various boundary conditions according to the FSDT and NET by means of the Ritz technique and the Bolotin's approach. Most recently, Huang et al. [8] applied an analytical solution to perform the dynamic stability of nanobeams exposed to an axially oscillating loading on the basis of the NET and EBT. Additionally, several works in literature have utilized the NET for the study of buckling, bending, vibration, and wave propagation problems of nanobeams/rods [9–12], nanoshells [13–15], and nanoplates [16–22].

It is recognized that in practical applications, nanostructures or nano reinforced composites often work in environmental conditions. Hence, the mechanical response of these structures at different environmental conditions has been a great subject of investigation among the scientific community. In this regard, Malekzadeh et al. [23] investigated the thermal buckling behavior of arbitrary straight-sided quadrilateral orthotropic nanoplates resting on Pasternak foundation by means of the nonlocal Kirchhoff plate theory and differential quadrature method (DQM). Ansari et al. [24] applied the NET within the framework of the EBT and the Timoshenko beam theory (TBT) to examine the dynamic stability of embedded SWCNTs in a thermal environment. In line with the previous works, Karličić et al. [25] analyzed the thermal vibration and buckling behavior of the embedded multi-layered graphene sheets (MLGSs) according to the NET and Kirchhoff–Love plate theory. Sobhy [26] applied state-space concept to carry out the static bending of Levy-type nanoplates subjected to hygrothermal and mechanical loads based on the NET in conjunction with two-variable plate theory. A DQM numerical approach was applied by Ebrahimi and Hosseini [27] to investigate the nonlinear vibration behavior of bilayer viscoelastic nanoplates in a thermal environment. Ansari and Gholami [28] studied the dynamic instability of multi-walled carbon nanotubes (MWCNTs) embedded on a Winkler medium including thermal effects, while applying a nonlocal TBT. The governing equations were solved via generalized DQM in conjunction with Bolotin's approach. Wu et al. [29] employed DQM and Bolotin's approach to explore the dynamic instability behavior of functionally graded graphene-reinforced nanocomposite plates under an axially oscillating loading and uniform temperature variation according to the FSDT. Jouneghani et al. [30] employed NET to study the bending characteristics of porous functionally-graded (FG) Timoshenko nanobeams exposed to hygro-thermo-mechanical loads. Further valuable research of nanostructures in a thermal environment can be found in [31–34].

Despite many works in literature have focused on the dynamic stability of nanobeams subjected to a thermal loading, there is a general lack of works focusing on the dynamic stability of nanoplates such as graphene sheets in a thermal environment. In most cases, the study of the thermal sensitivity of mechanical behavior of a SLGS is associated with temperature-independent material properties, whereas the experimental outcomes and MD-based simulations have proved the dependence of nanomaterial properties to temperature.

Hence, we propose a numerical study on the dynamic stability of a temperature-dependent SLGS embedded in a temperature-dependent elastomeric medium and subjected to a thermo-mechanical loading. This represents the key point of the present investigation, where the governing equations of motion are derived according to the FSDT in conjunction with the NET. These equations are then converted into a Mathieu–Hill type equation via the Navier's method, while obtaining the principle unstable region through the application of the Bolotin's method. A large systematic investigation evaluates the influences of the nonlocal parameter (NP), aspect ratio, temperature variation, foundation type, and static load factor on the dynamic instability region (DIR) of an orthotropic SLGS, as useful in design of nano electro-mechanical systems and micro electro-mechanical systems (NEMS/MEMS).

## 2. Basic Formulation

A temperature-dependent SLGS is here studied, with length $a$, width $b$, and thickness $h$ subjected to a biaxial periodic compressive loading in a thermal environment (Figure 1). The graphene sheet rests on an elastomeric temperature-dependent foundation, which is described by means of the spring constant $k_w$ and the shear layer parameter $k_p$. Among many possibilities of selecting the plate theory and solution methodology, in this work we select a FSDT, while considering possible shear effects on the structural response of the SLGS despite their atomic dimensions, as already demonstrated in [35]. The selected FSDT is here combined to the NET, whose basic notions are provided in the following overview.

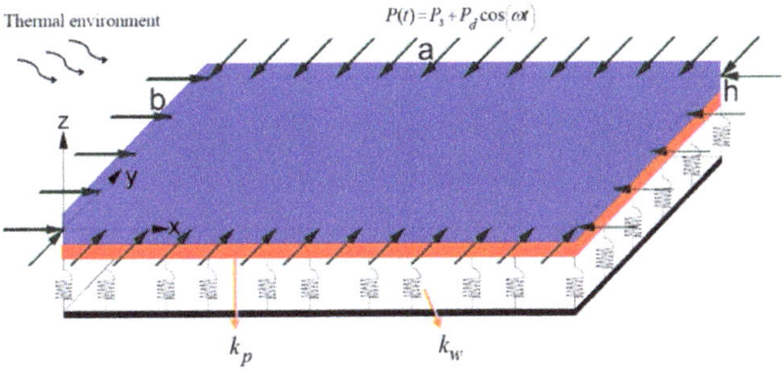


**Figure 1.** Configuration of the single-layered graphene sheet (SLGS) resting on an elastomeric foundation and subjected to a biaxial periodic compressive loading in a thermal environment.

### 2.1. Overview of NET

The NET introduced by Eringen [1], unlike the classical elasticity theory, states the stress tensor as a nonlocal function of the strain for the whole points in the domain. Hence, for a linear homogenous body, the nonlocal stress tensor at a point $x$ is given as:

$$\sigma_{ij}{}^{nl}(x) = \int_v \alpha\left(\left|x - x'\right|, \tau\right) \sigma_{ij}{}^{l} dV(x'), \qquad \forall x \in V \tag{1}$$

where $\sigma_{ij}^{nl}$ is the nonlocal stress, $\sigma_{ij}^{l}$ is the local stress tensors, $\alpha(|x - x'|, \tau)$ stands for the nonlocal kernel function. Additionally, $\tau$ is defined as $\tau = e_0 a/l$, where $a$ and $l$ represent the internal and external characteristic lengths, respectively, and the constant $e_0$ is introduced for adjusting the model according to experimental data or MD simulations. The differential constitutive form of Equation (1) can be written as follows [1,25], including the thermal effects, namely:

$$\left(1 - \mu\nabla^2\right)\sigma_{ij}^{nl} = \sigma_{ij}^{l} = C_{ijkl}(\varepsilon_{kl} - \alpha_{kl}\Delta T), \qquad \nabla^2 = \partial^2/\partial x^2 + \partial^2/\partial y^2. \tag{2}$$

In Equation (2), $\mu = (e_0 l)^2$ denotes the NP, $\alpha_{ij}$ is the thermal expansion coefficient, $\Delta T = T - T_0$ is the temperature variation with respect to the reference temperature $T_0$. In addition, $\varepsilon_{kl}$ indicates the strain tensor and $C_{ijkl}$ describes the fourth-order stiffness tensor.

Based on Equation (2), the constitutive relations based on the NET for the orthotropic nanoplate in thermal environment reads as follows:

$$\begin{Bmatrix} \sigma_{xx}^{nl} \\ \sigma_{yy}^{nl} \\ \sigma_{yz}^{nl} \\ \sigma_{xz}^{nl} \\ \sigma_{xy}^{nl} \end{Bmatrix} - \mu\,\nabla^2 \begin{Bmatrix} \sigma_{xx}^{nl} \\ \sigma_{yy}^{nl} \\ \sigma_{yz}^{nl} \\ \sigma_{xz}^{nl} \\ \sigma_{xy}^{nl} \end{Bmatrix} = \begin{bmatrix} C_{11} & C_{12} & 0 & 0 & 0 \\ C_{21} & C_{22} & 0 & 0 & 0 \\ 0 & 0 & C_{44} & 0 & 0 \\ 0 & 0 & 0 & C_{55} & 0 \\ 0 & 0 & 0 & 0 & C_{66} \end{bmatrix} \begin{Bmatrix} \varepsilon_{xx} - \alpha_{xx}\Delta T \\ \varepsilon_{yy} - \alpha_{yy}\Delta T \\ \gamma_{yz} \\ \gamma_{xz} \\ \gamma_{xy} \end{Bmatrix}. \quad (3)$$

in which the elastic constants of orthotropic materials can be determined by

$$C_{11} = \frac{E_1(T)}{1-\nu_{12}\nu_{21}}, \quad C_{12} = \frac{\nu_{12}E_2(T)}{1-\nu_{12}\nu_{21}}, \quad C_{22} = \frac{E_2(T)}{1-\nu_{12}\nu_{21}}, \quad C_{44} = G_{23}(T), \quad C_{55} = G_{13}(T), \quad C_{66} = G_{12}(T). \quad (4)$$

## 2.2. Strain Displacement Relationships

In what follows we apply the FSDT to formulate the governing equations of the problem, while including the thickness shear deformations and rotary effects. Thus, the displacement components associated to mid-surface $(u_0, v_0, w_0)$, transverse normal rotations $(\phi_x, \phi_y)$, and the displacement of an arbitrary point $(u, v, w)$ are related as [36]

$$\begin{aligned} u(x,y,z,t) &= u_0(x,y,t) + z\phi_x(x,y,t), \\ v(x,y,z,t) &= v_0(x,y,t) + z\phi_y(x,y,t), \\ w(x,y,z,t) &= w_0(x,y,t), \end{aligned} \quad (5)$$

in which $t$ was the time variable. The nonzero normal and transverse shear strains of the FSDT can be derived by

$$\begin{Bmatrix} \varepsilon_{xx} \\ \varepsilon_{yy} \\ \gamma_{xy} \end{Bmatrix} = \begin{Bmatrix} \varepsilon_{xx}^{(0)} \\ \varepsilon_{yy}^{(0)} \\ \gamma_{xy}^{(0)} \end{Bmatrix} + z \begin{Bmatrix} \varepsilon_{xx}^{(1)} \\ \varepsilon_{yy}^{(1)} \\ \gamma_{xy}^{(1)} \end{Bmatrix}, \quad \begin{Bmatrix} \gamma_{yz} \\ \gamma_{xz} \end{Bmatrix} = \begin{Bmatrix} \gamma_{yz}^{(0)} \\ \gamma_{xz}^{(0)} \end{Bmatrix}, \quad (6)$$

where

$$\begin{Bmatrix} \varepsilon_{xx}^{(0)} \\ \varepsilon_{yy}^{(0)} \\ \gamma_{xy}^{(0)} \end{Bmatrix} = \begin{Bmatrix} \frac{\partial u_0}{\partial x} \\ \frac{\partial v_0}{\partial y} \\ \frac{\partial u_0}{\partial y} + \frac{\partial v_0}{\partial x} \end{Bmatrix}, \quad \begin{Bmatrix} \varepsilon_{xx}^{(1)} \\ \varepsilon_{yy}^{(1)} \\ \gamma_{xy}^{(1)} \end{Bmatrix} = \begin{Bmatrix} \frac{\partial \phi_x}{\partial x} \\ \frac{\partial \phi_y}{\partial y} \\ \frac{\partial \phi_x}{\partial y} + \frac{\partial \phi_y}{\partial x} \end{Bmatrix}, \quad \begin{Bmatrix} \gamma_{yz}^{(0)} \\ \gamma_{xz}^{(0)} \end{Bmatrix} = \begin{Bmatrix} \phi_y + \frac{\partial w_0}{\partial y} \\ \phi_x + \frac{\partial w_0}{\partial x} \end{Bmatrix}. \quad (7)$$

# 3. Energy Method

## 3.1. Kinetic Energy

The kinetic energy for the nanoplate can be computed as

$$\begin{aligned} K &= \frac{\rho}{2}\int_A \int_{-h/2}^{h/2} \left( \left(\frac{\partial u}{\partial t}\right)^2 + \left(\frac{\partial v}{\partial t}\right)^2 + \left(\frac{\partial w}{\partial t}\right)^2 \right) \mathrm{d}A\mathrm{d}z \\ &= \frac{1}{2}\int_A \left\{ I_0 \left( \left(\frac{\partial u_0}{\partial t}\right)^2 + \left(\frac{\partial v_0}{\partial t}\right)^2 + \left(\frac{\partial w_0}{\partial t}\right)^2 \right) + 2I_1 \left( \left(\frac{\partial u_0}{\partial t}\frac{\partial \phi_x}{\partial t}\right) + \left(\frac{\partial v_0}{\partial t}\frac{\partial \phi_y}{\partial t}\right) \right) + I_2 \left( \left(\frac{\partial \phi_x}{\partial t}\right)^2 + \left(\frac{\partial \phi_y}{\partial t}\right)^2 \right) \right\} \mathrm{d}A \end{aligned} \quad (8)$$

where $\rho$ denotes the density of the orthotropic graphene sheet and $I_i$ refers to the mass inertias defined by

$$I_i = \rho \int_{-h/2}^{h/2} z^i \mathrm{d}z \quad (i = 0, 1, 2). \quad (9)$$

### 3.2. Strain Energy

The strain energy for the nanoplate could be obtained as

$$U = \frac{1}{2}\int_A \int_{-h/2}^{h/2} (\sigma_{xx}\varepsilon_{xx} + \sigma_{yy}\varepsilon_{yy} + K_s\sigma_{xz}\gamma_{xz} + K_s\sigma_{yz}\gamma_{yz} + \sigma_{xy}\gamma_{xy})\mathrm{d}A\mathrm{d}z. \tag{10}$$

In the above equation, $K_s$ stands for the shear correction factor that is here set to 5/6, as typically assumed in the literature for nanoplates and single-layer graphene sheets [20].

The combination of Equations (6), (7), and (10) yields

$$\begin{aligned} U = \frac{1}{2}\int_A & \left(N_{xx}\frac{\partial u_0}{\partial x} + M_{xx}\frac{\partial \phi_x}{\partial x} + N_{yy}\frac{\partial v_0}{\partial y} + M_{yy}\frac{\partial \phi_y}{\partial y} + N_{xy}\left(\frac{\partial u_0}{\partial y} + \frac{\partial v_0}{\partial x}\right)\right. \\ & \left. + M_{xy}\left(\frac{\partial \phi_x}{\partial y} + \frac{\partial \phi_y}{\partial x}\right) + Q_{xz}\left(\frac{\partial w_0}{\partial x} + \phi_x\right) + Q_{yz}\left(\frac{\partial w_0}{\partial y} + \phi_y\right)\right)\mathrm{d}A \end{aligned} \tag{11}$$

where

$$\left\{ \begin{array}{c} N_{\alpha\beta} \\ M_{\alpha\beta} \end{array} \right\} = \int_{-\frac{h}{2}}^{\frac{h}{2}} \sigma_{\alpha\beta}^{nl} \left\{ \begin{array}{c} 1 \\ z \end{array} \right\} \mathrm{d}z \qquad \alpha = x,\ y \quad \beta = x,\ y \tag{12}$$

$$Q_{\alpha z} = K_s \int_{-\frac{h}{2}}^{\frac{h}{2}} \sigma_{\alpha z}^{nl}\, \mathrm{d}z \qquad \alpha = x,\ y\,. \tag{13}$$

$N_{\alpha\beta}$ denotes the resultant force, $M_{\alpha\beta}$ describes the resultant moment, and $Q_{\alpha z}$ indicates the resultant shear force.

By substitution of Equations (3), (6), and (7) into Equations (12) and (13), the stress resultant-displacement relations are determined as

$$\begin{aligned} N_{xx} - \mu\nabla^2(N_{xx}) &= A_{11}\frac{\partial u_0}{\partial x} + A_{12}\frac{\partial v_0}{\partial y} - N_{xx}^T\,, \\ N_{yy} - \mu\nabla^2(N_{yy}) &= A_{12}\frac{\partial u_0}{\partial x} + A_{22}\frac{\partial v_0}{\partial y} - N_{yy}^T\,, \\ N_{xy} - \mu\nabla^2(N_{xy}) &= A_{66}\left(\frac{\partial u_0}{\partial y} + \frac{\partial v_0}{\partial x}\right), \end{aligned} \tag{14}$$

$$\begin{aligned} M_{xx} - \mu\nabla^2(M_{xx}) &= D_{11}\frac{\partial \phi_x}{\partial x} + D_{12}\frac{\partial \phi_y}{\partial y} - M_{xx}^T\,, \\ M_{yy} - \mu\nabla^2(M_{yy}) &= D_{12}\frac{\partial \phi_x}{\partial x} + D_{22}\frac{\partial \phi_y}{\partial y} - M_{yy}^T\,, \\ M_{xy} - \mu\nabla^2(M_{xy}) &= D_{66}\left(\frac{\partial \phi_x}{\partial y} + \frac{\partial \phi_y}{\partial x}\right), \end{aligned} \tag{15}$$

$$\begin{aligned} Q_{yz} - \mu\nabla^2(Q_{yz}) &= J_{44}\left(\phi_y + \frac{\partial w_0}{\partial y}\right), \\ Q_{xz} - \mu\nabla^2(Q_{xz}) &= J_{55}\left(\phi_x + \frac{\partial w_0}{\partial x}\right). \end{aligned} \tag{16}$$

More specifically, the extensional stiffness $A_{ij}$, the bending stiffness $D_{ij}$, and the shear stiffness $J_{ii}$ of the SLGS, are given as

$$\left(A_{ij},\ D_{ij}\right) = \int_{-h/2}^{h/2} C_{ij}\left(1,\ z^2\right)\mathrm{d}z \qquad (i,\ j = 1,2,6)\,, \tag{17}$$

$$J_{ii} = K_s \int_{-h/2}^{h/2} C_{ii}\,\mathrm{d}z \ = \ K_s h\, C_{ii} \qquad (i = 4,5)\,. \tag{18}$$

Additionally, $\left(N_{xx}^T, N_{yy}^T\right)$ and $\left(M_{xx}^T, M_{yy}^T\right)$ denote the thermally induced forces and moments due to a uniform temperature variation, respectively, determined by

$$\left\{ \begin{array}{c} N_{xx}^T \\ N_{yy}^T \end{array} \right\} = \int_{-\frac{h}{2}}^{\frac{h}{2}} \left\{ \begin{array}{c} A_{11}\alpha_{11} + A_{12}\alpha_{22} \\ A_{12}\alpha_{11} + A_{22}\alpha_{22} \end{array} \right\}\Delta T\ \mathrm{d}z, \tag{19}$$

$$\left\{\begin{array}{c} M_{xx}^{T} \\ M_{yy}^{T} \end{array}\right\} = \int_{-\frac{h}{2}}^{\frac{h}{2}} \left\{\begin{array}{c} A_{11}\alpha_{11} + A_{12}\alpha_{22} \\ A_{12}\alpha_{11} + A_{22}\alpha_{22} \end{array}\right\} \Delta T\, z\, \mathrm{d}z. \tag{20}$$

### 3.3. External Work

The external work done due to the temperature-dependent elastomeric medium and axially oscillating loading could be expressed as

$$V = \frac{1}{2} \int_A \left( -P_{xx} \frac{\partial^2 w_0}{\partial x^2} - P_{yy} \frac{\partial^2 w_0}{\partial y^2} - k_w w_0 + k_p \left( \frac{\partial^2 w_0}{\partial x^2} + \frac{\partial^2 w_0}{\partial y^2} \right) \right) w_0 \, \mathrm{d}A\,, \tag{21}$$

where $P_{xx} = P_{yy} = P$ denotes the axial dynamic forces. Additionally, the elastomeric foundation stiffness $k_w$ could be obtained as [37]

$$k_w = \frac{E_0}{4a(1 - \nu_0^2)(2 - c_1)^2} \left[ 5 - \left( 2\gamma_1^2 + 6\gamma_1 + 5 \right) \exp(-2\gamma_1) \right], \tag{22}$$

where

$$c_1 = (\gamma_1 + 2)\exp(-\gamma_1)\,, \quad \gamma_1 = \frac{H_s}{a}\,, \quad E_0 = \frac{E_s}{(1 - \nu_s^2)}\,, \quad \nu_0 = \frac{\nu_s}{1 - \nu_s}\,, \tag{23}$$

while $E_s$, $\nu_s$ and $H_s$ stand for the Young's modulus, Poisson's ratio, and depth of the foundation, respectively. In this research, $E_s$ is supposed to be temperature-dependent, whereas $\nu_s$ is kept constant.

## 4. Motion Equations

Hamilton's variational principle has been utilized in order to determine the equilibrium equations of nanoplate. Accordingly, one may write

$$\int_0^t (\delta U + \delta V - \delta K) dt = 0\,. \tag{24}$$

Inserting Equations (8), (11) and (21) into Equation (24) yields the following equations of motion

$$\delta u_0 : \quad \frac{\partial N_{xx}}{\partial x} + \frac{\partial N_{xy}}{\partial y} = I_0 \frac{\partial^2 u_0}{\partial t^2} + I_1 \frac{\partial^2 \phi_x}{\partial t^2}\,, \tag{25}$$

$$\delta v_0 : \quad \frac{\partial N_{xy}}{\partial x} + \frac{\partial N_{yy}}{\partial y} = I_0 \frac{\partial^2 v_0}{\partial t^2} + I_1 \frac{\partial^2 \phi_y}{\partial t^2}\,, \tag{26}$$

$$\delta w_0 : \quad \frac{\partial Q_{xz}}{\partial x} + \frac{\partial Q_{yz}}{\partial y} - (P_{xx} + N_{xx}^T) \frac{\partial^2 w_0}{\partial x^2} - \left( P_{yy} + N_{yy}^T \right) \frac{\partial^2 w_0}{\partial y^2} - k_w w_0 + k_p \left( \frac{\partial^2 w_0}{\partial x^2} + \frac{\partial^2 w_0}{\partial y^2} \right) = I_0 \frac{\partial^2 w_0}{\partial t^2}\,, \tag{27}$$

$$\delta \phi_x : \quad \frac{\partial M_{xx}}{\partial x} + \frac{\partial M_{xy}}{\partial y} - Q_{xz} = I_1 \frac{\partial^2 u_0}{\partial t^2} + I_2 \frac{\partial^2 \phi_x}{\partial t^2}\,, \tag{28}$$

$$\delta \phi_y : \quad \frac{\partial M_{xy}}{\partial x} + \frac{\partial M_{yy}}{\partial y} - Q_{yz} = I_1 \frac{\partial^2 v_0}{\partial t^2} + I_2 \frac{\partial^2 \phi_y}{\partial t^2}\,. \tag{29}$$

Finally, by substituting Equations (14)–(16) into Equations (25)–(29), the nonlocal governing equations of orthotropic graphene sheet resting on an elastomeric medium including the thermal effect can be determined as follows:

$$A_{11} \left( \frac{\partial^2 u_0}{\partial x^2} \right) + (A_{12} + A_{66}) \left( \frac{\partial^2 v_0}{\partial x \partial y} \right) + A_{66} \left( \frac{\partial^2 u_0}{\partial y^2} \right) + \left( 1 - \mu \nabla^2 \right) \left[ -I_0 \frac{\partial^2 u_0}{\partial t^2} - I_1 \frac{\partial^2 \phi_x}{\partial t^2} \right] = 0\,, \tag{30}$$

$$(A_{12} + A_{66})\left(\frac{\partial^2 u_0}{\partial x \partial y}\right) + A_{66}\left(\frac{\partial^2 v_0}{\partial x^2}\right) + A_{22}\left(\frac{\partial^2 v_0}{\partial y^2}\right) + \left(1 - \mu\nabla^2\right)\left[-I_0\frac{\partial^2 v_0}{\partial t^2} - I_1\frac{\partial^2 \phi_y}{\partial t^2 \partial y}\right] = 0\,, \quad (31)$$

$$\begin{array}{l} J_{55}\left(\frac{\partial^2 w_0}{\partial x^2} + \frac{\partial \phi_x}{\partial x}\right) \quad + J_{44}\left(\frac{\partial^2 w_0}{\partial y^2} + \frac{\partial \phi_y}{\partial y}\right) \\ \qquad + (1 - \mu\nabla^2)\left(\begin{array}{c} -k_w w_0 + k_p\left(\frac{\partial^2 w_0}{\partial x^2} + \frac{\partial^2 w_0}{\partial y^2}\right) - \left(P_{xx} + N_{xx}^T\right)\frac{\partial^2 w_0}{\partial x^2} \\ - \left(P_{yy} + N_{yy}^T\right)\frac{\partial^2 w_0}{\partial y^2} - I_0\frac{\partial^2 w_0}{\partial t^2} \end{array}\right) = 0\,, \end{array} \quad (32)$$

$$\begin{array}{l} D_{11}\left(\frac{\partial^2 \phi_x}{\partial x^2}\right) + D_{12}\left(\frac{\partial^2 \phi_y}{\partial x \partial y}\right) + D_{66}\left(\frac{\partial^2 \phi_x}{\partial y^2} + \frac{\partial^2 \phi_y}{\partial x \partial y}\right) - J_{55}\left(\frac{\partial w_0}{\partial x} + \phi_x\right) \\ \qquad + (1 - \mu\nabla^2)\left(-I_1\frac{\partial^2 u_0}{\partial t^2} - I_2\frac{\partial^2 \phi_x}{\partial t^2}\right) = 0\,, \end{array} \quad (33)$$

$$\begin{array}{l} D_{12}\left(\frac{\partial^2 \phi_x}{\partial x \partial y}\right) + D_{22}\left(\frac{\partial^2 \phi_y}{\partial y^2}\right) + D_{66}\left(\frac{\partial^2 \phi_x}{\partial x \partial y} + \frac{\partial^2 \phi_y}{\partial x^2}\right) - J_{44}\left(\frac{\partial w_0}{\partial y} + \phi_y\right) \\ \qquad + (1 - \mu\nabla^2)\left(-I_1\frac{\partial^2 v_0}{\partial t^2} - I_2\frac{\partial^2 \phi_y}{\partial t^2}\right) = 0\,. \end{array} \quad (34)$$

## 5. Solution Procedure

The governing differential equations are solved analytically via Navier's approach for simply-supported graphene sheet in this research. According to the selected approach, the mid-plane displacement and rotation field are solved by means of the Fourier series as

$$u_0(x, y, t) = \sum_{m=1}^{\infty}\sum_{n=1}^{\infty} U_{mn}(t)\cos(\alpha x)\sin(\beta y)\,, \quad (35)$$

$$v_0(x, y, t) = \sum_{m=1}^{\infty}\sum_{n=1}^{\infty} V_{mn}(t)\sin(\alpha x)\cos(\beta y), \quad (36)$$

$$w_0(x, y, t) = \sum_{m=1}^{\infty}\sum_{n=1}^{\infty} W_{mn}(t)\sin(\alpha x)\sin(\beta y), \quad (37)$$

$$\phi_x(x, y, t) = \sum_{m=1}^{\infty}\sum_{n=1}^{\infty} X_{mn}(t)\cos(\alpha x)\sin(\beta y), \quad (38)$$

$$\phi_y(x, y, t) = \sum_{m=1}^{\infty}\sum_{n=1}^{\infty} Y_{mn}(t)\sin(\alpha x)\cos(\beta y), \quad (39)$$

where $\alpha = \frac{m\,\pi}{a}$, $\beta = \frac{n\,\pi}{b}$, and $m$ is the half wave number in the $x$ direction and $n$ is the half wave number in the $y$ direction.

Finally, substituting Equations (35)–(39) into Equations (30)–(34), leads to the following matrix form, namely

$$\mathbf{M}\ddot{\mathbf{d}} + (\mathbf{K}_L - P(t)\mathbf{K}_G)\mathbf{d} = 0, \quad (40)$$

in which $\mathbf{d} = \{U \quad V \quad W \quad X \quad Y\}^T$ denotes the displacement vector. Additionally, $\mathbf{M}$ refers to the mass, $\mathbf{K}_L$ describes the stiffness, and $\mathbf{K}_G$ stands for the geometric stiffness matrices.

We consider a periodic compressive load which includes a static and dynamic part, as follows:

$$P(t) = P_s + P_d\cos(\omega t) = \alpha P_{cr} + \beta P_{cr}\;\cos(\omega t). \quad (41)$$

Here, $P_s$ refers to the static force component and $P_d$ indicates the dynamic force component, $P_{cr}$ shows the static buckling load and $\omega$ describes the pulsation frequency. Additionally, $\alpha$ is the static load factor and $\beta$ is the dynamic load factor. The substitution of Equation (41) into Equation (40) read

$$\mathbf{M}\ddot{\mathbf{d}} + (\mathbf{K}_L - (\alpha + \beta\cos(\omega t))P_{cr}\mathbf{K}_G)\mathbf{d} = 0\,. \quad (42)$$

Equation (42) is a Mathieu–Hill type equation discussing the dynamic instability behavior of graphene sheet under a periodic compressive load. The Bolotin's method [38] is employed to determine the boundaries of the DIR. Based on this method, the displacement vector of $d$ can be written in the following series with period $2T$ as

$$\mathbf{d} = \sum_{k=1,3,\ldots}^{\infty} \left[ a_k \sin\frac{k\,\omega\,t}{2} + b_k \cos\frac{k\,\omega\,t}{2} \right], \tag{43}$$

in which $a_k$ and $b_k$ denote arbitrary constant vectors. On the basis of Bolotin's approach [38], the first approximation with $k = 1$ could be used to compute the instability boundary. In this case, by setting Equation (43) into Equation (42) and considering a null value for the coefficients of the harmonic functions and the sum of the constant terms to zero, we get to the following relation:

$$\left| \left( \mathbf{K}_L - P_{cr}\left(\alpha \pm \frac{\beta}{2}\right)\mathbf{K}_G - \mathbf{M}\frac{\omega^2}{4} \right) \right| = 0\,. \tag{44}$$

At the end, we can compute the critical static buckling load, assuming $\beta = \omega = 0$. On the basis of the eigenvalue problem, the variation of $\omega$ versus $\beta$ can be plotted as instability regions of the nanoplate to the periodic load including thermal effect.

## 6. Numerical Examples

This section is devoted to the numerical study on the dynamic instability of orthotropic temperature-dependent SLGS resting on temperature-dependent elastomeric foundation undergoing biaxial oscillating loading in a thermal environment. Here, two types of zigzag SLGSs are considered with two different values of aspect ratio. The material properties of the SLGS are highly dependent to the temperature. Shen et al. [39] performed a MD simulation to evaluate these properties at three different temperatures, as listed in Table 1. In addition, a polydimethylsiloxane (PDMS) is here selected for the elastomeric foundation, while assuming the following material properties: a Poisson's ratio $\nu_s = 0.48$ and a Young's modulus $E_s = (3.22 - 0.0034T)\ GPa$, where $T = T_0 + \Delta T$ and $T_0 = 300\ K$ refers to the room temperature [37].

**Table 1.** Thermo-mechanical properties of the material [39].

| (K) | $E_1$ (TPa) | $E_2$ (TPa) | $G_{12}$ (TPa) | $\nu_{12}$ | $\alpha_{11}(10^{-6}$/K) | $\alpha_{22}(10^{-6}$/K) |
|---|---|---|---|---|---|---|
| | | | *Zigzag sheet I*: $a$ = 4.855 nm, $b$ = 4.888 nm, $h$ = 0.154 nm | | | |
| 300 | 1.987 | 1.974 | 0.857 | 0.205 | 2.1 | 1.9 |
| 500 | 1.974 | 1.968 | 0.870 | 0.205 | 2.3 | 2.0 |
| 700 | 1.961 | 1.948 | 0.870 | 0.205 | 2.4 | 2.1 |
| | | | *Zigzag sheet II*: $a$ = 9.496 nm, $b$ = 4.877 nm, $h$ = 0.145 nm | | | |
| 300 | 2.145 | 2.097 | 0.938 | 0.223 | 1.7 | 1.5 |
| 500 | 2.103 | 2.055 | 0.959 | 0.223 | 2.0 | 1.7 |
| 700 | 2.069 | 2.014 | 0.959 | 0.223 | 2.1 | 2.0 |

### 6.1. Validation of Results

A preliminary validation of the proposed mathematical formulation is performed against some results available in literature, see Tables 2 and 3. More specifically, Table 2 summarizes the results in terms of dimensionless natural frequencies for a simply-supported square graphene sheet, as computed by the present method, against predictions by Sobhy [40] based on SSDT, and or those ones by Ebrahimi and Barati [41] according to the higher-order refined plate theory and the DQM approach. A parametric investigation analyzes the variation of the NPs and foundation constants. The outcomes in Table 2 clearly show that the natural frequencies gradually decrease for an increased value of $\mu$ and increased for increasing values of $K_W$ and $K_P$. The very good agreement between our results and predictions from literature, clearly confirms the reliability of the present approach to capture the response of the problem, where few negligible differences are only related to the different selected solution methodology and plate theory.

**Table 2.** Dimensionless natural frequencies of the square graphene sheet for different nonlocal and foundation parameters ($a/h$ = 10).

| $\mu$ (nm$^2$) | $K_W$ | $K_P$ | Sobhy [40] | Ebrahimi and Barati [41] | Present | Error (%) |
|---|---|---|---|---|---|---|
| 0 | 0 | 0 | 1.93861 | 1.93861 | 1.93851 | 0.0051 |
| | 100 | 0 | 2.18396 | 2.18396 | 2.18386 | 0.0045 |
| | 100 | 20 | 2.96017 | - | 2.96010 | 0.0023 |
| 1 | 0 | 0 | 1.17816 | 1.17816 | 1.17809 | 0.0059 |
| | 100 | 0 | 1.54903 | 1.54903 | 1.54898 | 0.0032 |
| | 100 | 20 | 2.52831 | - | 2.52827 | 0.0015 |
| 2 | 0 | 0 | 0.92261 | 0.92261 | 0.92257 | 0.0043 |
| | 100 | 0 | 1.36479 | 1.36479 | 1.36475 | 0.0029 |
| | 100 | 20 | 2.41979 | - | 2.41976 | 0.0012 |
| 3 | 0 | 0 | 0.78347 | 0.78347 | 0.78343 | 0.0051 |
| | 100 | 0 | 1.27485 | 1.27485 | 1.27482 | 0.0023 |
| | 100 | 20 | 2.37020 | - | 2.37018 | 0.0008 |

**Table 3.** Comparative evlauation of the critical buckling load (nN/nm) for isotropic square graphene sheet under biaxial compression loads.

| $a$ (nm) | MD Results [42] | DQM [43] | FSM [19] | Present |
|---|---|---|---|---|
| 4.99 | 1.0837 | 1.0749 | - | 1.0711 |
| 8.08 | 0.6536 | 0.6523 | 0.6592 | 0.6514 |
| 10.77 | 0.4331 | 0.4356 | 0.4387 | 0.4353 |
| 14.65 | 0.2609 | 0.2645 | 0.2656 | 0.2643 |
| 18.51 | 0.1714 | 0.1751 | 0.1757 | 0.1751 |
| 26.22 | 0.0889 | 0.0917 | 0.0918 | 0.0917 |
| 33.85 | 0.0554 | 0.0561 | 0.0562 | 0.0561 |
| 37.81 | 0.0449 | 0.0453 | 0.0453 | 0.0452 |
| 45.66 | 0.0315 | 0.0313 | 0.0313 | 0.0313 |

As another attempt for validation, we compute the critical buckling loads for isotropic square graphene sheet under biaxial load with different lengths, whose results are summarized and compared in Table 3 with the ones based on an MD approach [42], a DQM approach [43], or a refined strip method (RFSM) [19]. The gradual increase of the geometrical dimension $a$ yields a meaningful reduction of the critical loading, whereas the very good agreement between our results and those reported in literature, confirms once again the accuracy of the proposed method.

### 6.2. Parametric Studies and Discussions

In the following examples, the DIR is traced as a dimensionless excitation frequency $\Omega = \omega b^2\sqrt{\rho h/D_{11}}$ in which $D_{11} = E_1 h^3/12(1-\nu_{12}\nu_{21})$ against the dynamic load factor $\beta$. It was supposed, unless otherwise explained, that $a/b = 1$ (zigzag sheet I) and $\mu = 0.5$ nm$^2$.

Figure 2 investigates the influence of the temperature change on the dynamic instability of orthotropic graphene sheet without elastomeric foundation. As clearly visible in Figure 2, an increase in temperature change reduces the excitation frequencies and shifts the origins of the DIR to the left side and also decreases gradually its width. The temperature rise leads to an increased compressive prestress in the graphene sheet, thus weakening the nanoplate stiffness.

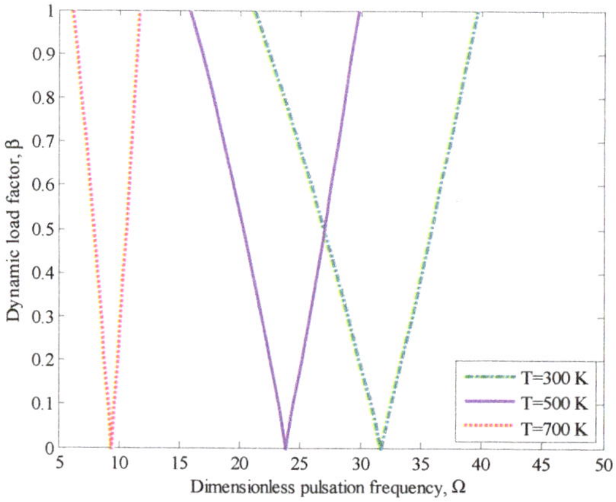


**Figure 2.** Temperature variation effect on the dynamic instability region (DIR) of the graphene sheet without elastomeric foundation.

As mentioned before, in this study we analyze the effect of the temperature on the mechanical properties of both the nanostructure and elastic foundation. Thus, Figure 3a,b plot the DIR of the SLGS resting on the elastomeric foundation under the twofold assumption of temperature-independent and temperature-dependent material properties. A comparative evaluation of the results is illustrated in Figure 3a,b, for two different values of temperature, i.e., $T = 500$ K and $T = 700$ K, respectively. As visible in Figure 3a,b, when temperature-dependent properties are used, the origins of the instability regions occur at smaller excitation frequencies. The difference between the two cases becomes even more pronounced as the temperature is enhanced. This means that it is significant to account for the effect of temperature on the material properties in order to obtain more reliable results.

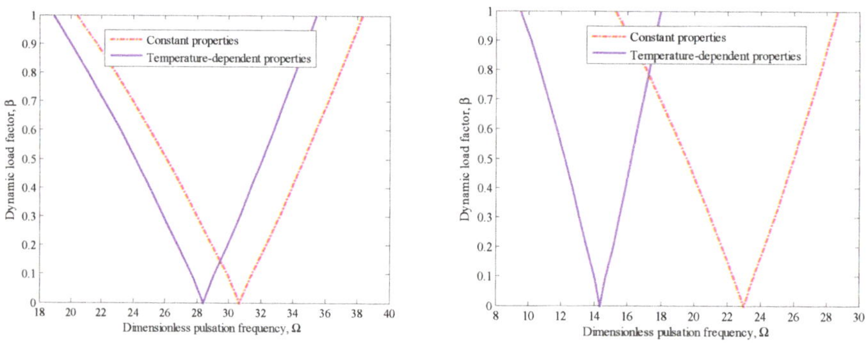


(a) $T = 500$ K (b) $T = 700$ K

**Figure 3.** Comparison between the DIR of SLGS resting on the elastomeric foundation for constant and temperature-dependent properties.

The surrounding elastic foundation effect on the DIR of SLGS is displayed in Figure 4a,b, for $T = 300$ K and $T = 700$ K, respectively. As visible in Figure 4a,b, the presence of an elastic foundation increases the excitation frequency and hence, the origin of unstable region moves to the right side. The existence of an elastomeric foundation magnifies the nanoplate stiffness which yields an increased stability of the nanoplate. In addition, by considering an elastic foundation, the instability region of the graphene sheet tends to become wider, especially at increased levels of temperature. Moreover, it can be observed that the effect of the elastomeric Pasternak medium ($K_W \neq 0$ and $K_P \neq 0$) is higher than the elastomeric Winkler ($K_W \neq 0$ and $K_P = 0$) one, on the DIR of the SLGS. The elastomeric Winkler foundation is able to describe only the normal loading condition of the elastomeric foundation, whereas the Pasternak model considers normal stresses as well as transverse shear loading conditions. From these figures, it is also worth observing that the foundation effects become even more significant for increased temperatures.

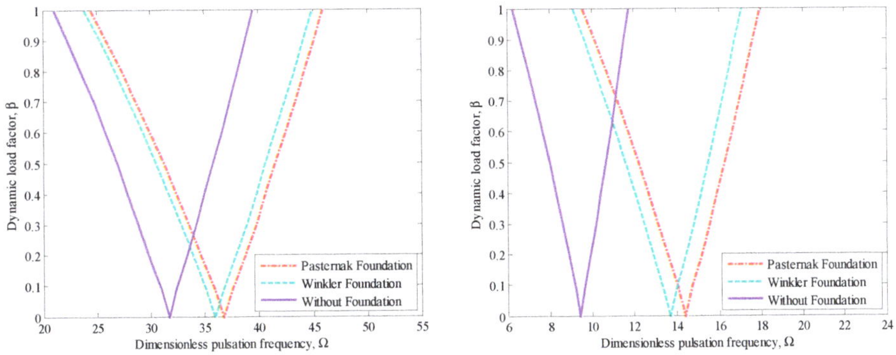


(**a**) $T = 300$ K (**b**) $T = 700$ K

**Figure 4.** Effect of the elastomeric foundation on the DIR of the SLGS.

Figure 5a,b illustrates the aspect ratio $a/b$ effect on the unstable region of graphene sheets embedded in the elastomeric foundation for $T = 300$ K and $T = 700$ K, respectively. In this example, two different values of aspect ratio (i.e., $a/b \simeq 1$ (zigzag sheet I) and $a/b \simeq 2$ (zigzag sheet II)) are considered [39]. This figure indicates that the excitation frequency reduces and the dynamic instability moves to the left side as $a/b$ increases. This trend is almost predictable, since the nanoplate becomes softer for increasing aspect ratios, with a global reduction of the structural stiffness. Moreover, the width of the instability region increases with a decrease in the slenderness ratio.

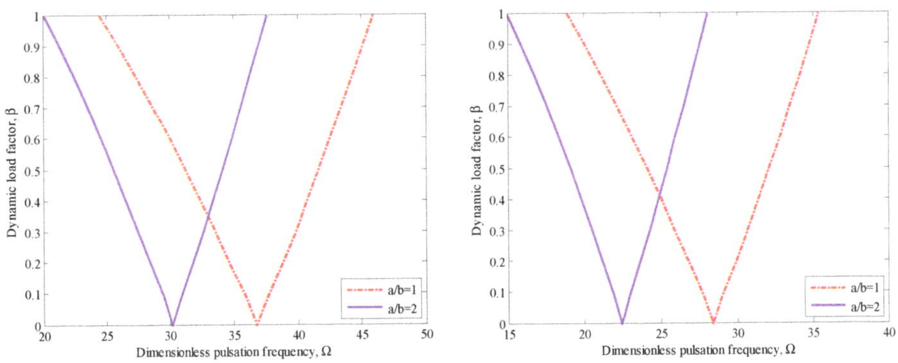


(**a**) $T = 300$ K (**b**) $T = 500$ K

**Figure 5.** Aspect ratio effect on the DIR of the SLGS embedded on an elastomeric foundation.

The NP effect on the DIR of embedded SLGS is displayed in Figure 6a,b, for $T = 300$ K and $T = 700$ K, respectively. The dimensionless pulsation frequency seems to decrease as the NP becomes higher, since the DIR of graphene sheet forms at lower excitation frequencies. Such phenomenon is owing to the fact that NP indicates the softening effect on stiffness which leads to lower pulsation frequency. Furthermore, comparing these figures exhibits that an increase in temperature change increases the effect of NP on the unstable region of the nanoplate.

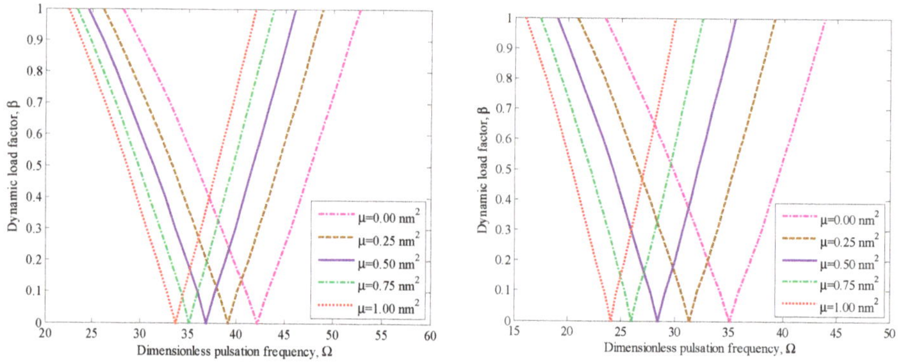


(**a**) $T = 300$ K (**b**) $T = 500$ K

**Figure 6.** Effect of the NP on the DIR of the SLGS resting on an elastomeric foundation.

Finally, Figure 7a,b demonstrates the static load factor effect on the unstable region of the SLGS resting on the elastomeric foundation for $T = 300$ K and $T = 500$ K, respectively. As can be concluded, with an increase in the static load factor the excitation frequency decreases and therefore, dynamic instability shifted to the left side due to the reduction in the stiffness of the nanoplate. Furthermore, the results indicate that the DIR becomes wider for a higher static load factor. Any static force component is present, in the particular case when $\alpha = 0$.

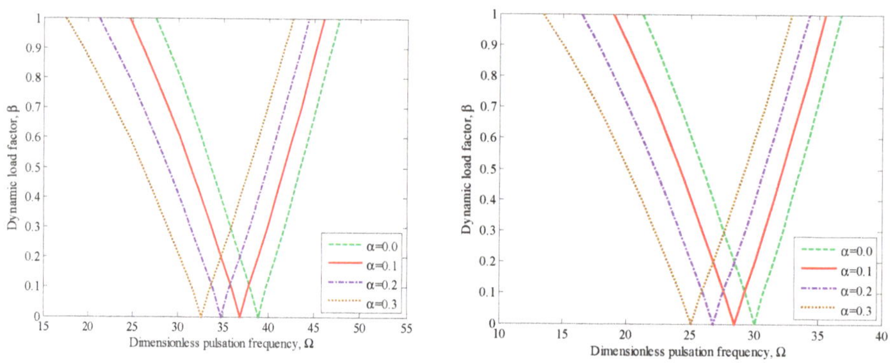


(**a**) $T = 300$ K (**b**) $T = 500$ k

**Figure 7.** Static load factor effect on the DIR of the SLGS embedded on an elastomeric foundation.

## 7. Conclusions

This study investigated the dynamic instability of temperature-dependent SLGS embedded in an elastic medium in thermal environment. This was modelled by means of a temperature-dependent elastomeric foundation. Based on FSDT within the framework of Eringen's differential constitutive model, the equations of motion were here derived from the Hamilton's variational principle as well as the energy method. The Navier's solution in conjunction with the Bolotin's approach were here applied to calculate the unstable region for the graphene sheet. A large numerical investigation examined the influence of different parameters such as the NP, temperature change, aspect ratio, static load factor, and foundation type on the dynamic stability behavior. Based on the numerical results, it was found that the elastic foundation increases the width of the instability region and the origin of the unstable region shift to the right side which leads to higher excitation frequencies. An increased static load factor and/or temperature weakens the nanoplate stiffness and leads to a lower pulsating frequency, while moving the DIR to the left side. It seemed also that the DIR occurs at smaller values of excitation frequency, when temperature-dependent mechanical properties were assumed for both the SLGS and the foundation, compared to the assumption of constant properties.

**Author Contributions:** Conceptualization, M.H.J., R.D. and F.T.; Formal analysis, M.H.J., R.D. and F.T.; Investigation, M.H.J. and F.T.; Validation, M.H.J., R.D. and F.T.; Writing—Original Draft, M.H.J., R.D. and F.T.; Writing—Review & Editing, R.D. and F.T.

**Funding:** This research received no external funding.

**Conflicts of Interest:** The authors declare no conflict of interest.

## References


1. Eringen, A.C. On differential equations of nonlocal elasticity and solutions of screw dislocation and surface waves. *J. Appl. Phys.* **1983**, *54*, 4703–4710. [CrossRef]
2. Murmu, T.; McCarthy, M.A.; Adhikari, S. In-plane magnetic field affected transverse vibration of embedded single-layer graphene sheets using equivalent nonlocal elasticity approach. *Compos. Struct.* **2013**, *96*, 57–63. [CrossRef]
3. Wang, Y.; Li, F.-M.; Wang, Y.-Z. Nonlinear vibration of double layered viscoelastic nanoplates based on nonlocal theory. *Phys. E Low-Dimens. Syst. Nanostruct.* **2015**, *67*, 65–76. [CrossRef]
4. Karličić, D.; Kozić, P.; Pavlović, R.; Nešić, N. Dynamic stability of single-walled carbon nanotube embedded in a viscoelastic medium under the influence of the axially harmonic load. *Compos. Struct.* **2017**, *162*, 227–243. [CrossRef]
5. Arani, A.G.; Jalaei, M.H. Investigation of the longitudinal magnetic field effect on dynamic response of viscoelastic graphene sheet based on sinusoidal shear deformation theory. *Phys. B Condens. Matter* **2017**, *506*, 94–104. [CrossRef]
6. Kolahchi, R.; Zarei, M.S.; Hajmohammad, M.H.; Naddaf-Oskouei, A. Visco-nonlocal-refined Zigzag theories for dynamic buckling of laminated nanoplates using differential cubature-Bolotin methods. *Thin-Walled Struct.* **2017**, *113*, 162–169. [CrossRef]
7. Jalaei, M.H.; Arani, A.G.; Tourang, H. On the dynamic stability of viscoelastic graphene sheets. *Int. J. Eng. Sci.* **2018**, *132*, 16–29. [CrossRef]
8. Huang, Y.; Fu, J.; Liu, A. Dynamic instability of Euler–Bernoulli nanobeams subject to parametric excitation. *Compos. Part B Eng.* **2019**, *164*, 226–234. [CrossRef]
9. Torabi, K.; Nafar-Dastgerdi, J. An analytical method for free vibration analysis of Timoshenko beam theory applied to cracked nanobeams using a nonlocal elasticity model. *Thin Solid Films* **2012**, *520*, 6595–6602. [CrossRef]
10. Adhikari, S.; Murmu, T.; McCarthy, M.A. Frequency domain analysis of nonlocal rods embedded in an elastic medium. *Phys. E Low-Dimens. Syst. Nanostruct.* **2014**, *59*, 33–40. [CrossRef]
11. Zhang, D.P.; Lei, Y.; Shen, Z.B. Vibration analysis of horn-shaped single-walled carbon nanotubes embedded in viscoelastic medium under a longitudinal magnetic field. *Int. J. Mech. Sci.* **2016**, *118*, 219–230. [CrossRef]
12. Wu, C.-P.; Lin, C.-H.; Wang, Y.-M. Nonlinear finite element analysis of a multiwalled carbon nanotube resting on a Pasternak foundation. *Mech. Adv. Mater. Struct.* **2018**, 1–13. [CrossRef]
13. Wang, Q.; Varadan, V.K. Application of nonlocal elastic shell theory in wave propagation analysis of carbon nanotubes. *Smart Mater Struct.* **2007**, *16*, 178. [CrossRef]
14. Shen, H.-S.; Zhang, C.-L. Torsional buckling and postbuckling of double-walled carbon nanotubes by nonlocal shear deformable shell model. *Compos. Struct.* **2010**, *92*, 1073–1084. [CrossRef]
15. Ansari, R.; Torabi, J. Numerical study on the free vibration of carbon nanocones resting on elastic foundation using nonlocal shell model. *Appl. Phys. A* **2016**, *122*, 1073. [CrossRef]
16. Pradhan, S.C.; Murmu, T. Small scale effect on the buckling of single-layered graphene sheets under biaxial compression via nonlocal continuum mechanics. *Comput. Mater. Sci.* **2009**, *47*, 268–274. [CrossRef]
17. Shi, J.-X.; Ni, Q.-Q.; Lei, X.-W.; Natsuki, T. Study on wave propagation characteristics of double-layer graphene sheets via nonlocal Mindlin–Reissner plate theory. *Int. J. Mech. Sci.* **2014**, *84*, 25–30. [CrossRef]
18. Arani, A.G.; Jalaei, M.H. Nonlocal dynamic response of embedded single-layered graphene sheet via analytical approach. *J. Eng. Math.* **2016**, *98*, 129–144. [CrossRef]

19. Sarrami-Foroushani, S.; Azhari, M. Nonlocal buckling and vibration analysis of thick rectangular nanoplates using finite strip method based on refined plate theory. *Acta Mech.* **2016**, *227*, 721–742. [CrossRef]
20. Arani, A.G.; Jalaei, M.H. Transient behavior of an orthotropic graphene sheet resting on orthotropic visco-Pasternak foundation. *Int. J. Eng. Sci.* **2016**, *103*, 97–113. [CrossRef]
21. Zhou, Z.; Rong, D.; Yang, C.; Xu, X. Rigorous vibration analysis of double-layered orthotropic nanoplate system. *Int. J. Mech. Sci.* **2017**, *123*, 84–93. [CrossRef]
22. Jalaei, M.H.; Ghorbanpour-Arani, A. Size-dependent static and dynamic responses of embedded double-layered graphene sheets under longitudinal magnetic field with arbitrary boundary conditions. *Compos. Part B Eng.* **2018**, *142*, 117–130. [CrossRef]
23. Malekzadeh, P.; Setoodeh, A.R.; Beni, A.A. Small scale effect on the thermal buckling of orthotropic arbitrary straight-sided quadrilateral nanoplates embedded in an elastic medium. *Compos. Struct.* **2011**, *93*, 2083–2089. [CrossRef]
24. Ansari, R.; Gholami, R.; Sahmani, S. On the dynamic stability of embedded single-walled carbon nanotubes including thermal environment effects. *Sci. Iran.* **2012**, *19*, 919–925. [CrossRef]
25. Karličić, D.; Cajić, M.; Kozić, P.; Pavlović, I. Temperature effects on the vibration and stability behavior of multi-layered graphene sheets embedded in an elastic medium. *Compos. Struct.* **2015**, *131*, 672–681. [CrossRef]
26. Sobhy, M. Hygrothermal deformation of orthotropic nanoplates based on the state-space concept. *Compos. Part B Eng.* **2015**, *79*, 224–235. [CrossRef]
27. Ebrahimi, F.; Hosseini, S.H.S. Thermal effects on nonlinear vibration behavior of viscoelastic nanosize plates. *J. Therm. Stress.* **2016**, *39*, 606–625. [CrossRef]
28. Ansari, R.; Gholami, R. Dynamic stability analysis of multi-walled carbon nanotubes with arbitrary boundary conditions based on the nonlocal elasticity theory. *Mech. Adv. Mater. Struct.* **2017**, *24*, 1180–1188. [CrossRef]
29. Wu, H.; Yang, J.; Kitipornchai, S. Parametric instability of thermo-mechanically loaded functionally graded graphene reinforced nanocomposite plates. *Int. J. Mech. Sci.* **2018**, *135*, 431–440. [CrossRef]
30. Jouneghani, F.Z.; Dimitri, R.; Tornabene, F. Structural response of porous FG nanobeams under hygro-thermo-mechanical loadings. *Compos. Part B Eng.* **2018**, *152*, 71–78. [CrossRef]
31. Wang, Y.-Z.; Li, F.-M.; Kishimoto, K. Thermal effects on vibration properties of double-layered nanoplates at small scales. *Compos. Part B Eng.* **2011**, *42*, 1311–1317. [CrossRef]
32. Zenkour, A.M. Nonlocal transient thermal analysis of a single-layered graphene sheet embedded in viscoelastic medium. *Phys. E Low-Dimens. Syst. Nanostruct.* **2016**, *79*, 87–97. [CrossRef]
33. Zhang, D.P.; Lei, Y.J.; Shen, Z.B. Thermo-electro-mechanical vibration analysis of piezoelectric nanoplates resting on viscoelastic foundation with various boundary conditions. *Int. J. Mech. Sci.* **2017**, *131–132*, 1001–1015. [CrossRef]
34. Malikan, M.; Nguyen, V.B.; Tornabene, F. Damped forced vibration analysis of single-walled carbon nanotubes resting on viscoelastic foundation in thermal environment using nonlocal strain gradient theory. *Eng. Sci. Technol. Int. J.* **2018**, *21*, 778–786. [CrossRef]
35. Ebrahimi, F.; Shafiei, N. Influence of initial shear stress on the vibration behavior of single-layered grapheme sheets embedded in an elastic medium based on Reddy's higher-order shear deformation plate theory. *Mech. Adv. Mater. Struct.* **2016**. [CrossRef]
36. Jalaei, M.H.; Arani, A.G. Analytical solution for static and dynamic analysis of magnetically affected viscoelastic orthotropic double-layered graphene sheets resting on viscoelastic foundation. *Phys. B Condens. Matter* **2018**, *530*, 222–235. [CrossRef]
37. Shen, H.-S.; Zhang, C.-L. Nonlocal beam model for nonlinear analysis of carbon nanotubes on elastomeric substrates. *Comput. Mater. Sci.* **2011**, *50*, 1022–1029. [CrossRef]
38. Bolotin, V.V. *The Dynamic Stability of Elastic Systems*; Holden-Day: San Francisco, CA, USA, 1964.
39. Shen, L.; Shen, H.-S.; Zhang, C.-L. Temperature-dependent elastic properties of single layer graphene sheets. *Mater. Des.* **2010**, *31*, 4445–4449. [CrossRef]
40. Sobhy, M. Thermomechanical bending and free vibration of single-layered graphene sheets embedded in an elastic medium. *Phys. E Low-Dimens. Syst. Nanostruct.* **2014**, *56*, 400–409. [CrossRef]
41. Ebrahimi, F.; Barati, M.R. Damping vibration analysis of graphene sheets on viscoelastic medium incorporating hygro-thermal effects employing nonlocal strain gradient theory. *Compos. Struct.* **2018**, *185*, 241–253. [CrossRef]

42. Ansari, R.; Sahmani, S. Prediction of biaxial buckling behavior of single-layered graphene sheets based on nonlocal plate models and molecular dynamics simulations. *Appl. Math. Model.* **2013**, *37*, 7338–7351. [CrossRef]
43. Golmakani, M.E.; Rezatalab, J. Nonuniform biaxial buckling of orthotropic nanoplates embedded in an elastic medium based on nonlocal Mindlin plate theory. *Compos. Struct.* **2015**, *119*, 238–250. [CrossRef]

*applied sciences*

MDPI

*Article*

# Nonlocal and Size-Dependent Dielectric Function for Plasmonic Nanoparticles

**Kai-Jian Huang [1,2], Shui-Jie Qin [1,2,*], Zheng-Ping Zhang [1], Zhao Ding [1] and Zhong-Chen Bai [2]**

1 College of Big Data and Information Engineering, Guizhou University, Guiyang 550025, China
2 Guizhou Provincial Key Lab for Photoelectron Technology and Application, Guizhou University, Guiyang 550025, China
* Correspondence: shuijie_qin@sina.com

Received: 4 July 2019; Accepted: 26 July 2019; Published: 31 July 2019

**Abstract:** We develop a theoretical approach to investigate the impact that nonlocal and finite-size effects have on the dielectric response of plasmonic nanostructures. Through simulations, comprehensive comparisons of the electron energy loss spectroscopy (EELS) and the optical performance are discussed for a gold spherical dimer system in terms of different dielectric models. Our study offers a paradigm of high efficiency compatible dielectric theoretical framework for accounting the metallic nanoparticles behavior combining local, nonlocal and size-dependent effects in broader energy and size ranges. The results of accurate analysis and simulation for these effects unveil the weight and the evolution of both surface and bulk plasmons vibrational mechanisms, which are important for further understanding the electrodynamics properties of structures at the nanoscale. Particularly, our method can be extended to other plasmonic nanostructures where quantum-size or strongly interacting effects are likely to play an important role.

**Keywords:** EELS; plasmons vibrational modes; nanoparticles; nonlocal and size-dependent dielectric

## 1. Introduction

Classical local electrodynamics in nanostructures under the assumption that the polarization characteristics at a given point are locally related to the external electromagnetic excitation, which has been thoroughly explored from isolated nanoparticles and dimers to other complex structures [1–5]. On the other hand, the nonlocal response of a metal nanostructure to an external disturbance stimulus determines many important plasmonic properties and has attracted much theoretical interest [6–9]. In recent years, the Mortensen group [10] have studied plasmons in metal nanoparticles by using a popular hydrodynamic model which incorporates nonlocal effects by introducing a pressure term that inhibits squashing electrons into small plasmonic nanoparticles. Similarly, the work of Christensen et al. [11] included nonlocality in the optical response of metallic nanoparticles by introducing quantum corrections obtained from first principles calculations, also using the hydrodynamic model. Ming et al. [12,13] have demonstrated a Maxwell-hydrodynamic model for studying the charge, energy and angular momentum conservation laws of high-order harmonics due to nonlocal and nonlinear interactions between electromagnetic waves and metallic metamaterials in the terahertz regime.

Meanwhile, parallel to the above remarkable progress of nonlocal properties, there has been a growing interest in exploring the quantum properties of surface plasmon and the prospects of plasma devices based on reliable operation at the quantum level. However, theoretical investigation of plasmonic resonances for nanostructure sizes ranging between 2 and 20 nm remains a challenge. This length scale is too small for classical electromagnetic theories to be valid but too large for full-fledged quantum simulations to be feasible. Therefore, some of the studies are committed to provide a unified theoretical framework that breaks wide discrepancy between classical electromagnetic theory and quantum mechanical theory for plasmonics and nanophotonics, and ultimately control plasmonic responses in nanostructures. Crucial to this endeavor is to understand finite-size effects in nanostructures since the control of nanoparticle size represents one of the most effective approaches to harness the tremendous potentials of plasmonics and nanophotonics.

Recent studies have shown a theoretical method with a suitable modification of the bulk complex dielectric function for free and bound electrons within metallic nanoparticles [14–19], for the determination of plasma frequency, damping constants, and size distributions of noble metal nanoparticles, such as gold, silver and copper; in addition, the corresponding optical characterization deduced by Mie's theory has been tested to fit full UV-Vis experimental extinction spectra. The above theoretical framework follows the idea of Kreibig and von Fragstein [20], who considered that the augmentation of the damping constant in the Drude model results from additional collisions of free electrons with the particle boundary, as well as the idea of Inouye et al. [21], who took into account the interband transitions of bound electrons for the complex dielectric function. Consequently, the metal dielectric function can be regarded as the sum of the free (intraband transitions) and bound (interband transitions) electron contributions; then, a size-dependent (SD) dielectric function (Equation (4)) based on a stepwise modification of the Drude model is put forward. Under this theoretical framework, the significant finite-size effects of optical properties for metallic nanoparticles can be expected to be observed below a certain limited size range (e.g., valid restriction of 20 nm drawn from Refs. [14,15]).

A developed microscopic approach based on accurate inclusion of the Lorentz friction for the plasmon damping within the random phase approximation (RPA) model [22–24] has also been performed in recent years. Unlike the conventional Mie's theory that only gives reasonably intrinsic size effects for nanospheres of radii lower than 5 nm in vacuum due to neglecting the Lorentz friction losses that dominate over all attenuation channels, the specific size dependence for plasmons due to the Lorentz friction induced corrections derived upon this theoretical approach can be applied to account for the observed irregular size-effect of plasmon resonance for relatively large metallic nanoparticles (e.g., for Au nanospheres of radii 20–60 nm in vacuum drawn from Ref. [24]). This theoretical approach has also been verified to improve experimental data fitting of metallic nanospheres measured in water colloidal solutions, especially pronounced for a large size limit (e.g., of radii 10–75 nm for Au nanospheres in water colloidal solutions drawn from Ref. [22]). One should note that, although accurate accounting for the Lorentz friction is important for nanospheres of radii above 20 nm (for Au in vacuum), the major drawback of our work is that adequate corrections of the Lorentz friction are not included, since integrating them with the nonlocal and size-dependent effects within a macroscopic material dielectric function is not straightforward.

Extending beyond the nonlocal and size-dependent effects we have focused on, in the following, there is a huge amount of research concerning about other "non-classical electromagnetic effects" by incorporating some "quantum effects" within classical electrodynamics. Quantum-size effects due to the confinement of electron motion in small dimensions can lead to a discrete electronic energy levels and a size-dependent conductivity [25]. Accessing plasmonic modes in the quantum tunneling regime also remains as a challenge, quantum tunneling effects become significant for subnanometer interparticle gaps and can be reflected through a quantum corrected model (QCM), which models the junction between adjacent nanoparticles by means of a local dielectric response that includes electron tunneling and tunneling resistivity at gap [26,27]. However, such "non-classical" effects are of great interest and significance in the field of nano-optics, and can be incorporated into classical electrodynamical descriptions for some simple systems; directly integrating them within a general framework is not straightforward. In particular, quantum tunneling effects between metal nanoparticles will not be considered in this work.

Nonlocal effects are a direct consequence of the inhomogeneity of the electron gas, and can hardly be ignored when the mean free path of an electron in the simple systems is larger than the wavelength of light [28,29]. To a certain extent, nonlocal electrodynamics partially account for finite-size and quantum mechanical effects inherent in a nanostructure in a more rigorous way. Such as in the case of applying the hydrodynamic dielectric function to calculate the absorption coefficient from metallic nanoparticles, its size-dependent effects will show up by an effective dielectric function due to the additional boundary conditions for solving polarizabilities [30]. Although it has been argued that the hydrodynamic model yields only semiquantitative results in the linear regime, recent improvements regarding the incorporation of interband transitions and Landau damping have demonstrated that fully quantitative results may be obtained [31,32]. In principle, as most of the modified hydrodynamic models shown in the aforementioned applications, a spatially nonlocal relationship between the material polarization and the field of plasmonics can typically be mirrored through the use of a spatially nonlocal dielectric function. Some of them take finite-size effects of the nanostructure into consideration by using a modified collision frequency. However, the corresponding changes in optical, plasmonic and dielectric responses are almost negligible, and the effective dimension of nanostructure for nonlocal electrodynamics is also being limited or obscured consequently.

Inspired by the studies of the size-dependent dielectric function as mentioned above, in this paper, we move a step forward in this direction by developing nonlocal and size-dependent (NLSD) dielectric approach that incorporates the spatially local, nonlocal, size and even analogous quantum-size responses of the material. As an example, we investigate how the electron energy loss spectroscopy (EELS) and optical responses of gold spherical dimers of various radii and inter-particle gaps are modified once the nonlocality and finite-size effects are considered. These examples allow for a straightforward interpretation of our results. Unlike those complex configurational and resource consuming methods, such as quantum mechanical simulations based on the time-dependent Kohn–Sham density functional theory (TD-KSDFT), our approach provides a powerful and simple tool to calculate and analyze the EELS and optical response ranging from classical to quantum regimes.

## 2. Methods and Theoretical Framework

Figure 1 displays the fundamental system under study; a dimer consists of two closely spaced freestanding metallic nanospheres of radius $R$, and a certain distance $d$ apart. The system is excited by an electron beam with high kinetic energy up to 200 keV passing by or penetrating through at the positions as indicated in the figure, and illuminated from the top by a plane wave polarized along the dimer axis, respectively. For this study, the metal is assumed to be gold, and the experimental and theoretical values of permittivity for contrast taken from Johnson and Christy [33] and Werner et al. [34], respectively. We chose the data models above for comparison for the following reasons. First, of all, in general, ab initio density functional theory (DFT) approaches need to deal with pseudopotentials, and include quantum effects that completely change the spectral distribution and the field enhancements of the plasmon resonant modes supported by a given nanostructure; they are computationally demanding and are currently limited to small-sized nanoparticles with a few thousand electrons (e.g., valid restriction of 12.3 nm drawn from Ref. [35]). Second, the application of classic local experimental dielectric data for small-sized nanoparticles ranging between 2 and 20 nm becomes questionable as described in the Introduction. These factors will be discussed in detail below combined with calculation results. All of the numerical simulations are implemented via the general public license toolbox MNPBEM [36,37], which is based on a boundary element method (BEM) approach.

The dielectric function of noble metal nanoparticles (e.g., Cu, Au, and Ag) is well described by using the hydrodynamical Drude model limit by three components [38–40]:

$$\varepsilon_{\text{Nonlocal}}(\mathbf{k},\omega) = \varepsilon_{\infty} + \varepsilon_{\text{inter}}(\omega) + \varepsilon_{\text{intra}}(\mathbf{k},\omega), \quad (1)$$

where the $\mathbf{k}$ dependence in $\varepsilon_{\text{Nonlocal}}(\mathbf{k},\omega)$ is necessary to describe the spatially nonlocal response of structures with features less than 10 nm [41,42]. $\varepsilon_{\infty}$ is the value as $\omega \to \infty$. $\varepsilon_{\text{inter}}(\omega)$ represents the contribution from interband electron transitions and can be physically described in the form of a Lorentz oscillator model,

$$\varepsilon_{\text{inter}}(\omega) = \sum_j \frac{\Delta\varepsilon_{\text{L}j}\omega_{\text{L}j}^2}{\omega_{\text{L}j}^2 - \omega(\omega + i2\delta_{\text{L}j})}. \tag{2}$$

In this work, we take index $j = 2$ labeling the individual d-band to sp-band electron transitions since there are two interband transitions in Au at optical frequencies [33]. $\varepsilon_{\text{intra}}(\mathbf{k},\omega)$ is responsible for the intraband effects which are usually referred to as free-electron effects, where both the plasmonic optical response of metals and nonlocal effects are contained and can be described by

$$\varepsilon_{\text{intra}}(\mathbf{k},\omega) = -\frac{\omega_{\text{D}}^2}{\omega(\omega + i\gamma) - \beta^2\mathbf{k}^2}, \tag{3}$$

where $\omega_{\text{D}}$ is the plasma frequency, $\gamma$ is the collision frequency, and $\beta$ is the hydrodynamical parameter measuring the degree of nonlocality. For the three-dimensional situation $\beta^2 = 3v_{\text{F}}/5$, where $v_{\text{F}} = 1.39 \times 10^6$ m/s is the Fermi velocity for the Au nanoparticles. Then, all parameters in Equations (2) and (3) can be estimated by fitting Equation (1) to the experimentally determined dielectric data of bulk Au in Ref. [33]. Here, we use all the parameters taken from Ref. [39], $\omega_{\text{D}} = 8.812$ eV, $\gamma = 0.0752$ eV, $\varepsilon_\infty = 3.559$, $\Delta\varepsilon_{\text{L1}} = 2.912$, $\omega_{\text{L}j} = 4.693$ eV, $\delta_{\text{L1}} = 1.541$ eV, $\Delta\varepsilon_{\text{L2}} = 1.272$, $\omega_{\text{L}j} = 3.112$ eV, $\delta_{\text{L1}} = 0.525$ eV.

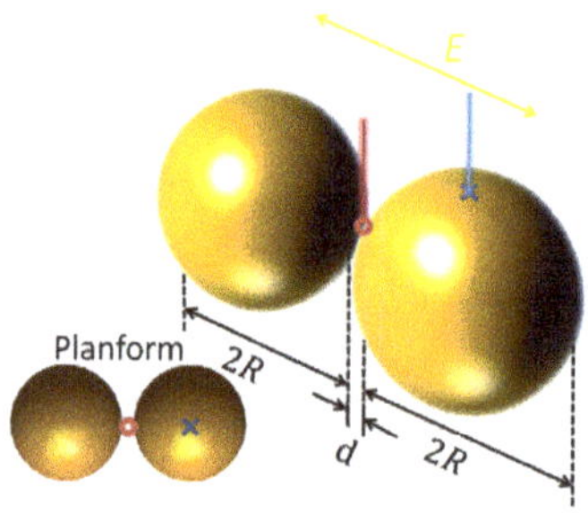


**Figure 1.** Schematic illustration of the system under study: the interaction of the electron beam passes by or penetrates through a dimer consisted with two Au nanospheres with radius R and gap d, as well as the plane wave polarized along the dimer axis. The inset shows the exact located positions (or impact parameters) of electron beams: the red ring represents the middle of the gap and the blue cross represents the centre in one of the nanospheres.

Note that the above dielectric function calculation model neglects quantum-mechanical exchange and correlation effects, but the significant nonlocal effects from the reduced mean free path of the conduction electrons due to electron–interface scattering in small metal structures can be taken into account by using a modified collision frequency $\gamma' = \gamma + 4Av_{\text{F}}V/S$ in Equation (3) for the three-dimensional problems [43]. In principle, conclusions from the previews studies that adopt such modified collision frequency have shown that the Nonlocal model of Equation (1) can model the spatially nonlocal dielectric response of arbitrarily shaped and sized structures [39,40]. However, although the volume of the structure $V$ and surface area $S$ in the modified collision frequency implies the influence of size factor, only the impact of free electrons was considered to be modified by size. In addition, correctly selecting the value of dimensionless parameter $A$ remains a challenging since its complex details are of discrepancy in different theoretical dielectric models (consult Section 1 in Ref. [44] and its references for a detailed discussion). Therefore, in order to overcome this difficulty in a classical mean free path approach, a more reasonable alternative way for describing size-dependent dielectric functions is needed by considering the dominated conditions and a scope of the contributions of intraband transitions (free electrons) and interband transitions (bound electrons). Herein, according to previous studies about suitable modification of the bulk complex dielectric function for free and bound electrons within metallic nanoparticles [14–16], the full size-corrected dielectric function can be written as the sum of three terms:

$$\varepsilon_{\mathrm{SD}}(\omega, R) = \varepsilon_{\mathrm{bulk}}(\omega) + \Delta\varepsilon_{\mathrm{free}}(\omega, R) + \Delta\varepsilon_{\mathrm{bound}}(\omega, R), \tag{4}$$

where $\varepsilon_{\mathrm{bulk}}(\omega)$ corresponding to the experimentally measured bulk values from UV-visible to NIR-FIR, and the last two terms are size-corrective contributions for free and bound electrons, respectively:

$$\Delta\varepsilon_{\mathrm{free}}(\omega, R) = -\omega_{\mathrm{D}}^2 \sum_{n=1}^{\infty} (-1)^n \frac{(i\omega C v_{\mathrm{F}}/R)^n}{(\omega^2 + i\omega\gamma_{\mathrm{free}})^{n+1}}, \tag{5}$$

$$\begin{aligned} \Delta\varepsilon_{\mathrm{bound}}(\omega, R) = &- K_b e^{R/R_0} \int_{\omega_g}^{\infty} \frac{\sqrt{x-\omega_g}}{x} [1 - F(x,T)] \\ &\times \frac{dx}{x^2 - \omega^2 + \gamma_{\mathrm{bound}} - 2i\omega\gamma_{\mathrm{bound}}}, \end{aligned} \tag{6}$$

where $\hbar\omega_g$ is the gap energy; $F(x,T)$ is the Fermi distribution function of conduction electron of energy $\hbar x$ and temperature $T$ with Fermi energy $E_{\mathrm{F}}$, $\gamma_{\mathrm{free}} = \gamma$, $\gamma_{\mathrm{bound}}$ stands for the damping constant in the band to band transition; $K_b$ is a proportionality factor with units of $\mathrm{s}^{-3/2}$; $C$ is a constant that depends on the material. For noble metals, it has a value of about 0.8 that has been theoretically justified from first principles calculations [45]; $R_0$ is a reference radius value that represents the range for which the density of states can be considered to reach the value of the bulk. For Au, the following values determined in Refs. [14–16] were used: $\omega_g = 2.1062$ eV, $\gamma_{\mathrm{bound}} = 0.158$ eV, $K_b = 2.3 \times 10^{24}\ \mathrm{s}^{-3/2}$, $E_{\mathrm{F}} = 2.5$ eV.

Note that the interband transitions from the d-band to the conduction sp-band near the L point in the Brillouin zone can be taken into account by the size-corrective contribution for bound electrons in the form of Equation (6) [46]. Because of the nonlocal dielectric function, $\varepsilon_{\mathrm{Nonlocal}}(\mathbf{k},\omega)$ is estimated by fitting Equation (1) to the complex bulk dielectric function, which is the local, frequency-dependent part of the response, taken from optical measurements given in Ref. [33] and exactly the same as the first term $\varepsilon_{\mathrm{bulk}}(\omega)$ using the size-dependent dielectric function $\varepsilon_{\mathrm{SD}}(\omega, R)$ of Equation (4); we adopt here the former as a substitute for the latter. The resulting nonlocal and size-dependent (NLSD) dielectric function of the metal is thus approximated by

$$\varepsilon_{\mathrm{NLSD}}(\mathbf{k}, \omega, R) = \varepsilon_{\mathrm{Nonlocal}}(\mathbf{k},\omega) + \Delta\varepsilon_{\mathrm{free}}(\omega, R) + \Delta\varepsilon_{\mathrm{bound}}(\omega, R). \tag{7}$$

## 3. Results and Discussion

The first example of the compatible effectiveness produced by our NLSD approach is shown in Figure 2 for comparisons of EELS calculated with different dielectric models and gold spherical dimers of different sizes, where electron beam penetrates the centre in one of the nanospheres (Figure 2a) and passes by the middle of the gap (Figure 2b) as shown in Figure 1, and the gap $d$ fixed on 1 nm. Figure 2a shows that the agreement of the total energy loss probabilities between the NLSD approach(black solid line) and the theoretical dielectric data set (green dashed line), which is obtained from density functional theory (DFT) calculations by Werner et al. [34], is remarkable for radii up to 20 nm. Moreover, for the NLSD approach, the evolves of the size-dependent response as the radius increases are observably more outstanding than the full size-dependent dielectric model (blue dotted line, Equation (4)), and the total energy loss probabilities become almost consistent with those calculated by using the local, frequency-dependent and experimental complex bulk dielectric values (red dash-dotted line) as radii up to 100 nm.

In addition, blueshifts of the main energy loss peaks become more apparent as the radius increases and it is accompanied by plasmon broadening. These nonlocal effects are known as the availability of the additional loss channels [7], and can have an effect in the higher energy region when compared to classical local electrodynamics because of the interplay between $\omega$ and $\mathbf{k}$ in Equation (3). All of the above agreements between the model predictions support that our approach provides a straightforward theoretical framework to describe metal behavior combining local, nonlocal and size-dependent effects in broader energy and size ranges simultaneously.

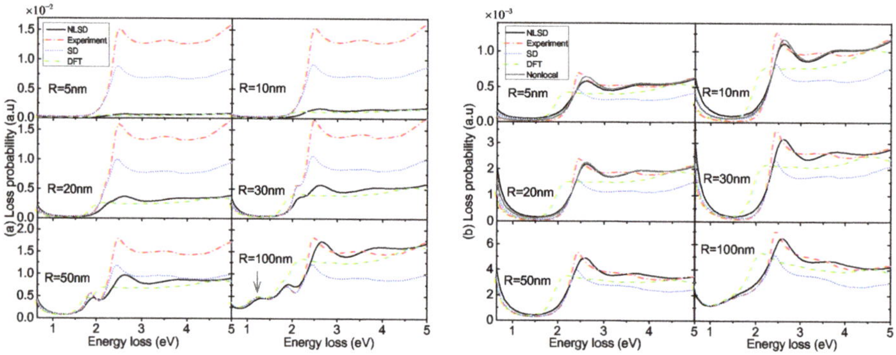


**Figure 2.** EELS (electron energy loss spectroscopy) spectrums for Au spherical dimers of different radii $R$, and for different dielectric models (black solid curves, NLSD (nonlocal and size-dependent) model; blue dotted curves, SD model; red dash-dotted curves, experimental dielectric data from Johnson et al. [33]; green dashed curves, DFT (density functional theory) calculated data from Werner et al. [34]; thinner gray solid curves, Nonlocal model), where the gap $d$ fixed on 1 nm. (**a**) electron beam penetrating the centre in one of the nanospheres, and (**b**) passing by the middle of the gap, respectively, as indicated in Figure 1.

Moreover, according to the various developed theories of the SD model as mentioned in Section 1, the significant finite-size effects on the optical properties of nanoparticles can only be expected to be observed below a certain limited size range of about 20 nm. Our model captures the main aspects of the finite-size effects for small-sized nanoparticles. For radii below 20 nm, Figure 2b shows a significant difference of the total (and surface) energy loss probabilities between the NLSD approach and nonlocal dielectric model (thinner gray solid curves) of Equation (1). $A = 0.1$ was used in the modified collision frequency $\gamma' = \gamma + 4Av_F V/S$ when using the Nonlocal model [39,40]. In addition, unlike the outcomes between the SD model and the experimental dielectric data, which show remarkable differences in all size range, the total (and surface) energy loss probabilities calculated by using the NLSD model are almost identical with the Nonlocal model for radii larger than 20 nm this is why the forecasted results of Nonlocal model are not shown in Figure 2b where the radii larger than 20 nm. Similarly, for clarity, Figure 2a does not show the outcomes from the Nonlocal model because they are identical with those from the NLSD model in all size ranges.

In consideration of the size-corrective terms from SD and NLSD model being the same, these results might appear unexpected at the first view. The nonlocal effects themselves are also size-dependent. This can be seen from Figure 2b where radii are below 20 nm; besides capturing the blueshifts, the main energy loss peaks of the Nonlocal model are obviously less than those by using experimental dielectric data, but in a low-energy region (e.g., 0.65–2.4 eV), the outcomes of the Nonlocal model are higher than those by using experimental dielectric data, which have the same changing tendency as the results of SD and NLSD models. However, the NLSD model is a more suitable implementation for revealing the contribution of the surface vibrational mode with nanoparticles of the small radius. When compared with the Nonlocal model, the outcomes of the NLSD show that the relative reductions in the main energy loss peaks, and the increases of the outcomes in the low-energy region are more remarkable, as shown in Figure 2b. All of these are because the size-dependent effects due to the size-corrective terms are contained in the NLSD model.

For further confirmation of the present findings, EELS calculated with different dielectric models for isolated Au spherical nanoparticles of different sizes are presented in Figure 3, where electron beam passing near the nanoparticles and the contribution of the surface vibrational mode is dominant. The difference of the total (and surface) energy loss probabilities between the NLSD model and Nonlocal model becomes more pronounced with the reduction of particle size, as shown in Figure 3 where the radius $R = 1$ nm, and the prediction of our NLSD model are much more close to the outcomes from the SD model and the DFT model. The discrepancy in predicted energy loss probabilities between the NLSD model and the Nonlocal model would gradually narrow and ultimately disappear when the radius is greater than 26 nm, as shown in Figure 3b, in addition to the discrepancy in predicted energy loss probabilities between the SD model and experimental dielectric data. These results for isolated nanoparticles are more compatible with the conclusions that have been verified both experimentally and theoretically in the previous studies of the SD model in which the significant size-dependent effects can only be expected to be observed below a certain limited size range of approximately 20 nm.

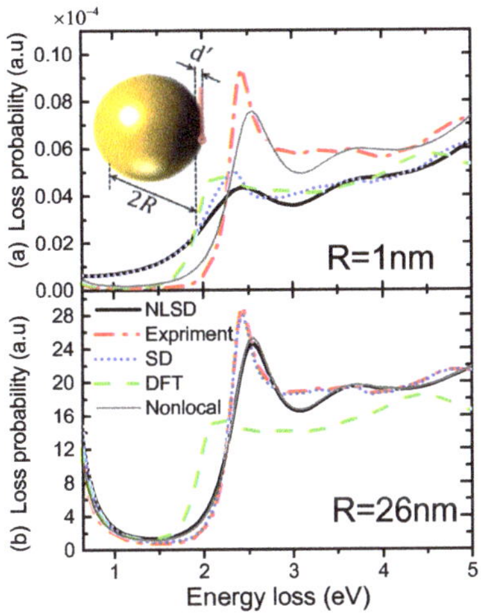


**Figure 3.** EELS spectrums for isolated Au spherical nanoparticles of different radii $R$, and for different dielectric models as defined in the text, when electron beam passing near the nanoparticles as depicted in the inset. Distances $d'$ between electron beam and nanoparticles were fixed at 2 nm.

Although, under what specific physical mechanism will the differences due to the size-dependent effects between the SD model and experimental dielectric data disappear in the outcomes between the NLSD and Nonlocal model for dimers of a large radius merit further investigation. However, according to all of the above results of EELS, one can see that the NLSD model presented in this paper has the advantage of clearly separating the size regions where local, nonlocal or size-dependent effects are dominant, for both simulation cases (isolated particle or dimer).

First, the size-dependent effects somehow can be revealed in small isolated nanoparticles by using the SD model, whereas the outcomes of this model do not contain any appreciable impact of the nonlocal effects. In addition, at least, for dimers of a large radius, the outcomes of the SD model should be much more consistent with those by using the classic local experimental dielectric data—particularly for an electron beam penetrating the centre in one of the nanospheres and the contribution of the bulk vibrational mode being dominant, as shown in Figure 2a. By contrast, the size-dependent effects due to the same size-corrective terms in the NLSD model have been demonstrated that, while they can be completely neglected for Au dimers of all size when the contribution of the bulk vibrational mode is dominant (Figure 2a), and, for Au dimers of a large radius, when the contribution of the surface vibrational mode is dominant (Figure 2b), they still have a great influence on dimers of a small radius, just as the results have been demonstrated for isolated particles (Figure 3).

Second, we should notice that the classic local experimental dielectric data used in this paper was obtained from reflection and transmission measurements on vacuum-evaporated thin films at room temperature [33]; the application of such local continuum electrodynamics for small-sized nanoparticles becomes questionable. This can be attributed to two effects. First of all, their imaginary part of the interband contribution due to size-dependent effects is not negligible in the spectral region of surface plasmon polariton resonance [14,15]. Second, nonlocal effects due to the scattering of the conduction electrons arise as the nanoparticle size is decreased, relative to metal films [39,40]. By contrast, either the EELS results or the optical properties that are going to be demonstrated in the following have shown that outcomes from our NLSD model can fit the DFT model better than other models for small-sized nanoparticles. These agreements benefit from the first term in the NLSD model, namely, the Nonlocal model of Equation (1), which have the advantage of describing the dynamical optical response of structures that are too large to treat using quantum mechanics but too small for classical electromagnetic theories to be valid [39,40].

It should be noted that a consistent energy loss with peak near 1.25 eV when the radius $R = 100$ nm (as indicated by the inset gray arrow in Figures 2a and 4(upper)), which is similar to the main plasmon resonances, will be presented below in Figure 6a for the extinction cross sections excited by optical plane wave, from approximately 955 nm. In order to gain physical insight into the origin of these common resonances in both situations of excitations, the surface and bulk configurations that form the total loss probabilities where $R = 100$ nm are plotted in Figure 4, respectively. The results shown in Figure 4 indicate that the total loss probabilities near 1.25 eV for any dielectric model are all dominated by surface loss probabilities. In addition, it is interesting that we found that the total energy loss probabilities shown in Figure 2b are also dominated by the surface loss probabilities, since all the bulk loss probabilities are almost zero when the electron beam located in the middle of the gap (not shown). We attribute this surface energy loss dominating effect to the contribution of the surface vibrational mode.

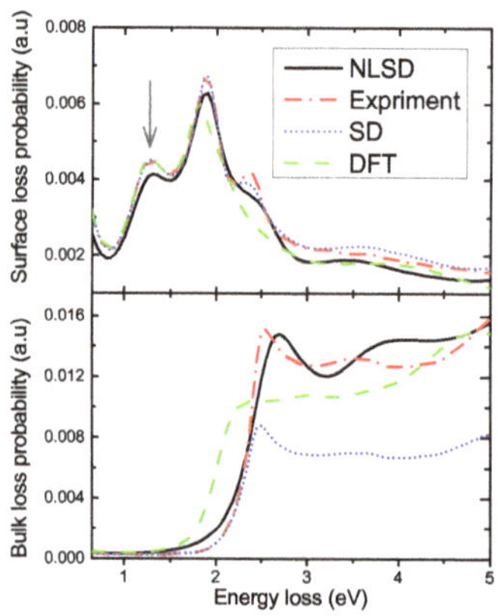


**Figure 4.** (Upper panel) Surface and (lower panel) bulk loss probabilities for gold spherical dimers of radius $R = 100$ nm, where the other simulation configurations remain the same as defined in Figure 2a.

One should note that, at least for now, the main energy loss peaks as depicted in Figure 2a are the coupling contribution of the surface and the bulk vibrational modes. However, it can be seen that these surface energy loss peaks near 1.25 eV as shown in Figure 4(upper) do not show up in

Figure 2b where the contribution of the surface vibrational mode is dominant. In addition, according to our surface loss probability calculation results (not shown, similar to the form of Figure 4(upper)), they also do not show up when $R = 50$ nm where the other simulation configurations remain the same as defined in Figure 2a. However, the reducing contribution of the surface vibrational mode with nanoparticles of the small radius can be distinctly revealed by the NLSD model. For example, Figure 5 plots the simulated EELS probability maps of our dimer system with four different dielectric models for the 1.25 eV resonances. Our model captures the main aspects of the nonlocal electrodynamics for small-sized nanoparticles.

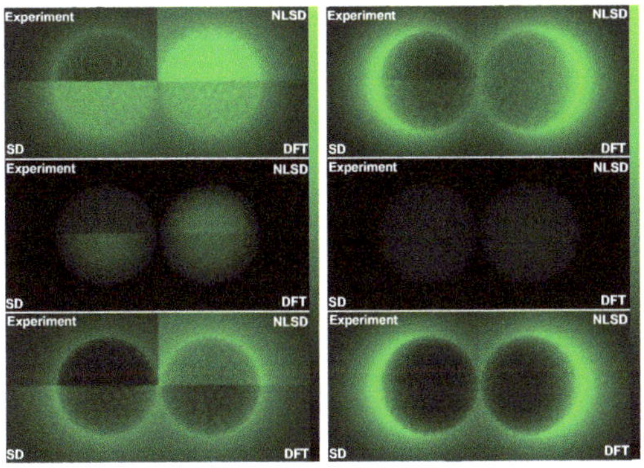


**Figure 5.** Total (top panels), bulk (middle panels) and surface (bottom panels) EELS probability maps for the gold dimers with particles radii $R = 20$ nm (left column) and $R = 50$ nm (right column), and for different dielectric models as defined in the text. The inter-particle gap $d = 1$ nm.

For radius $R = 20$ nm, the surface and bulk vibrational modes by using the NLSD model are higher than the others as shown in the middle and bottom panel of Figure 5 (left column). Thus, the energy loss extended throughout the whole surface of the gold spherical dimer, whereas those by using experimental dielectric data are confined in narrow rings formed by the surface and the horizontal cross-section of the dimer, as the total EELS probabilities shown in the left-top panel in Figure 5. We found similar results for dimers with radii less than 20 nm, except that the intensities of the energy loss calculated by using the NLSD approach the DFT model as the radius decreases.

For radius $R = 50$ nm, the total energy loss confined more strongly in narrow rings formed by the surface and the horizontal cross-section of the dimer for all dielectric models, and they are dominated by surface vibrational mode, as indicated by the EELS probability maps shown in Figure 5 (right column). We also found similar results for dimers with radii greater than 50 nm, except that the patterns of the total (and surface) energy loss changing from the narrow rings to crescents as the radius increases. These results mean that the nonlocal impact on the mode evolution of EELS and responses from the quantum mechanical mechanism can also be depicted by using our NLSD model. Although there are similar outcomes of mode evolution by using the SD and DFT models, they are obviously incompatible with the classic local experimental data where the size of the dimer becomes larger, such as shown in Figure 2 (e.g., $R = 100$ nm).

Finally, we use our NLSD approach to explore another interesting compatible effectiveness for the optical properties of the dimer system illuminated by a plane wave as defined in Figure 1. Although the effects of quantum natural and nonlocality have different origins, their impact on the optical response of the nanostructure can be measured by its physical parameters (in our case, which means the radius and gap). Figure 6a renders the maximum plasmonic resonances of extinction cross-sections that correspond to different radius for gold spherical dimers with the constant gap $d = 1$ nm, and for different dielectric models. The comparison between predictions from our NLSD model and the other dielectric models clarifies how nonlocality and quantum mechanical effects modify the optical properties of the system.

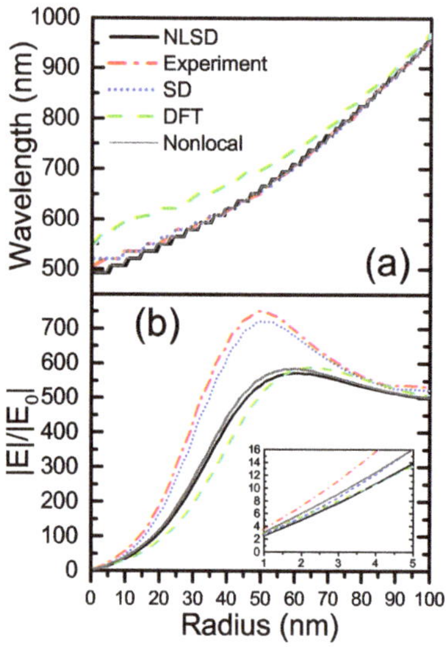


**Figure 6.** (**a**) maximum plasmonic resonances of extinction cross-sections that correspond to different radius for gold spherical dimers with constant inter-particle gap $d = 1$ nm, and for different dielectric models as defined in text; (**b**) maximum field enhancement at the middle point of the inter-particle gap (the red circle depicted in Figure 1) as a function of $R$ for gold spherical dimers and their corresponding plasmonic resonances as shown in (**a**).

For small radii (approximately less than 35 nm), there are significant blueshifts for the prediction of our NLSD model when compared to the predictions of SD model and experimental dielectric data, which are similar to the comparison results by using another entirely different nonlocal dielectric model of the metal from previous studies (see Figure 2 or Figure 4a in Ref. [7]), and can be accounted for the inclusion of nonlocal effects in the NLSD model, which also means that the size-dependent effects due to the size-corrective terms in the NLSD model can not affect the maximum plasmonic resonances of extinction cross-sections, since the predictions of our NLSD model and the SD model are approximately consistent with those of the Nonlocal model and experimental dielectric data, respectively, as shown in Figure 6a. One should note that these maximum plasmonic resonances of extinction cross-sections that correspond to different radius for gold spherical dimers of radii ranging between 20 and 60 nm may be more accurate if our model can be integrated with the adequate corrections of the Lorentz friction. However, nevertheless, either our NLSD model or the RPA approach included with the Lorentz friction induced corrections as discussed in Section 1 has observed anomalous shift of plasmonic resonance for nanoparticles with radii significantly exceeding the previously suggested limiting 5 nm by Mie's theory.

Figure 6b renders the field enhancement $\frac{|E|}{|E_0|}$ at the middle point of the gap as a function of $R$ for gold spherical dimers and their corresponding plasmonic resonances as shown in Figure 6a. Quite different from the extinction spectral profiles predicted, Figure 6b shows approximate tendency between the NLSD model and the DFT model. This means that the impacts of nonlocality and quantum confinement can be compatible, and both reduce the enhancement of fields. The quantum confinement effects are important when the mean free path of valence electrons is comparable to the particle size, and the mean free path of electrons in gold at room temperature is below 50 nm. These play a large role in interpreting the discrepancy in predicted field enhancements between the NLSD model and experimental dielectric data would gradually narrow when the radius is greater than 50 nm, as shown in Figure 6b. Similar behavior occurs for the prediction of the DFT model, except that the radius which stands out for such change is approximately 65 nm. We attribute this difference to the inclusion of nonlocal and finite-size effects in the NLSD model, which lead to an increase in the contribution of the bulk vibrational mode and to reduce the contribution of the surface vibrational mode when the radius becomes less than 50 nm and gradually decreases, as the EELS results (Figure 5) discussed above. Both effects are not exclusive of small particles as the inset shown in Figure 6b.

In addition, they are also important for small inter-particle gaps since the strongly interacting between metallic geometries encountered in plasmonics. For example, as one can see from the maximum field enhancement as a function of separation for gold dimer with $R = 50$ nm shown in Figure 7, the predictions of the NLSD model and the DFT model are closely similar, and the gaps in predicted field enhancements between them and the SD model and the experimental dielectric data would gradually narrow or even vanish as the inter-particle gap $d$ increases. As we can see both in Figures 6b and 7, between NLSD and nonlocal models, there is also a slight difference in their outcomes with a radius of less than approximately 65 nm (Figure 6b), or with a gap of less than approximately 0.5 nm (Figure 7). These are entirely caused by the size-dependent effects due to the size-corrective terms of the NLSD model, which also lead to a decreasing trend in the outcomes of the NLSD model toward the DFT model, particularly for dimers with small radii as the inset shown in Figure 6b. Furthermore, we also should note that the maximum field enhancement shown in Figure 7 may be not very accurate when the inter-particle gap $d$ decreases below 0.5 nm, since the NLSD model may not be capable of calculating the quantum tunneling effects under such subnanometer gaps, as mentioned in Section 1.

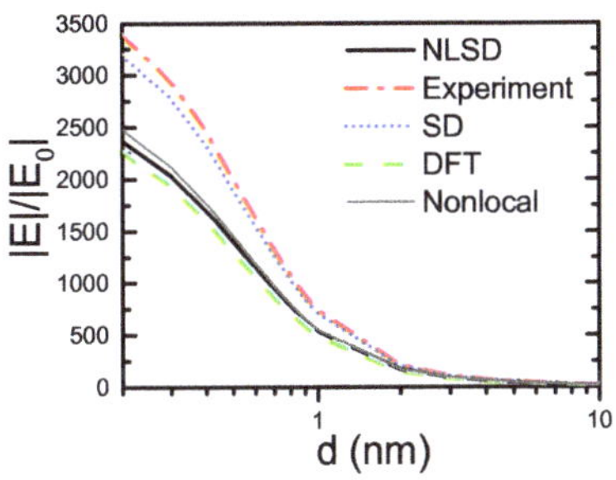


**Figure 7.** Maximum field enhancement as a function of separation for gold dimer with $R = 50$ nm, and for different dielectric models as defined in text (and depicted in Figure 2).

## 4. Conclusions

In conclusion, we presented a simplified theoretical model to describe the nonlocal and size effects of nanoparticles. Our implementation is based on the hydrodynamic Drude model and two size-corrective terms derived from the size-dependent dielectric function. We have demonstrated the compatible effectiveness and the accuracy of our theoretical model through comparisons to analytical results of the EELS and the optical responses for gold spherical dimers. These analytical results demonstrated a number of effects that result from nonlocality and finite-size effects in the dielectric response, including analogous quantum confinement effects that decrease the electromagnetic field enhancement and their corresponding weight relations of the contributions of surface and bulk vibrational modes reflected by EELS, blueshifts of surface plasmon resonances and main energy loss peaks. From these results, we can state that the nonlocal effects are of great significance as they obviously determine the blueshifts of surface plasmon resonances for the gold dimers with radii smaller than approximately 35 nm and the electromagnetic field enhancement for the gold dimers with radii ranging from 20 to 50 nm. In addition, blending the adequate size-dependent effects and nonlocal effects can more accurately reflect reductions in the main energy loss peaks and the electromagnetic field enhancement due to quantum confinement effects for the gold dimers with radii smaller than approximately 20 nm, as well as the weight relations of the contributions of surface and bulk vibrational modes for the gold dimers with radii ranging from 1 to 50 nm; for the gold dimers with radii larger than 50 nm, the nonlocal effects in the EELS and optical responses decrease as the radii increase further, and the local effects almost dominate the EELS and optical responses as radii up to 100 nm.

Our NLSD model used here accounts for the plasmonic nanoparticles behavior combining local, nonlocal and size-dependent effects, and reconciles the discrepancy between classical electromagnetic theory and quantum mechanical theory for plasmonics and nanophotonics in broader energy and size ranges. Importantly, although we have focused on gold spherical dimers, our approach can be extended to motivate new, and more precise experimental studies on other metallic nanostructures with complex configurations, particularly those where quantum-size effects and strongly interacting effects are likely to play a large role. In the future, we expect that our approach will make possible a deep exploration of the dissipative quantum electromagnetics, where couple loss or dissipation in a quantum system to a bath of oscillators maintained a strong connection with the classical electromagnetic system [47].

**Author Contributions:** Conceptualization, K.-J.H.; Formal analysis, K.-J.H.; Funding acquisition, Z.-C.B.; Methodology, K.-J.H.; Supervision, S.-J.Q.; Writing—original draft, K.-J.H.; Writing—review and editing, S.-J.Q., Z.-P.Z., Z.D. and Z.-C.B.

**Funding:** This research was funded by the National Natural Science Foundation of China (Grant No. 61741505) and the Guizhou Province Science and Technology Projects (No. QKHZ [2017]2887).

**Conflicts of Interest:** The authors declare no conflict of interest.

## References

1. Encai, H.; Schatz, G.C. Electromagnetic fields around silver nanoparticles and Dimers. *J. Chem. Phys.* **2004**, *120*, 357–366.
2. Amir, R.S.; Suhaidi, S.; Yap, W.F. Nanoplasmonic Sensor Based on Surface Plasmon-Coupled Emission: Review. *Appl. Sci.* **2019**, *9*, 1497.
3. Jacak, J.; Jacak, W. Plasmons and Plasmon–Polaritons in Finite Ionic Systems: Toward Soft-Plasmonics of Confined Electrolyte Structures. *Appl. Sci.* **2019**, *9*, 1159. [CrossRef]
4. Huigao, D.; Fernández-Domínguez, A.I.; Michel, B.; Maier, S.A.; Yang, J.K.W. Nanoplasmonics: Classical down to the nanometer scale. *Nano Lett.* **2012**, *12*, 1683–1689.
5. Xu, H. Theoretical study of coated spherical metallic nanoparticles for single-molecule surface-enhanced spectroscopy. *Appl. Phys. Lett.* **2004**, *85*, 5980–5982. [CrossRef]
6. David, C.; Abajo, F.J.G.D. Spatial Nonlocality in the Optical Response of Metal Nanoparticles. *J. Phys. Chem. C* **2011**, *115*, 19470–19475. [CrossRef]
7. Abajo, F.J.G.D. Non-local effects in the plasmons of strongly interacting nanoparticles, dimers, and waveguides. *J. Phys. Chem. C* **2008**, *112*, 17983–17987. [CrossRef]
8. Esquivel-Sirvent, R.; Schatz, G.C. Spatial Nonlocality in the Calculation of Hamaker Coefficients. *J. Phys. Chem. C* **2012**, *116*, 420–424. [CrossRef]
9. Moeferdt, M.; Kiel, T.; Sproll, T.; Intravaia, F.; Busch, K. Plasmonic modes in nanowire dimers: A study based on the hydrodynamic Drude model including nonlocal and nonlinear effects. *Phys. Rev. B* **2018**, *97*, 075431. [CrossRef]
10. Raza, S.; Bozhevolnyi, S.I.; Wubs, M.; Mortensen, N.A. Nonlocal optical response in metallic nanostructures. *J. Phys. Condens. Matter* **2015**, *27*, 183204. [CrossRef]
11. Christensen, T.; Yan, W.; Jauho, A.P.; Soljačić, M.; Mortensen, N.A. Quantum corrections in nanoplasmonics: Shape, scale, and material. *Phys. Rev. Lett* **2017**, *118*, 157402. [CrossRef] [PubMed]
12. Fang, M.; Huang, Z.X.; Sha, W.E.I.; Xiong, X.Y.; Wu, X. Full hydrodynamic model of nonlinear electromagnetic response in metallic metamaterials. *Prog. Electromagn. Res.* **2016**, *157*, 63–78. [CrossRef]
13. Fang, M.; Shen, N.H.; Sha, W.E.I.; Huang, Z.X.; Koschny, T.; Soukoulis, C.M. Nonlinearity in the Dark: Broadband Terahertz Generation with Extremely High Efficiency. *Phys. Rev. Lett.* **2019**, *122*, 027401. [CrossRef] [PubMed]

14. Herrera, L.J.M.; Arboleda, D.M.; Schinca, D.C.; Scaffardi, L.B. Determination of plasma frequency, damping constant, and size distribution from the complex dielectric function of noble metal nanoparticles. *J. Appl. Phys.* **2014**, *116*, 233105. [CrossRef]
15. Herrera, L.J.M.; Arboleda, D.M.; Santillán, J.M.J.; Raap, M.B.F.V.; Scaffardi, L.B.; Schinca, D.C. Nanoscale Dielectric Function of Fe, Pt, Ti, Ta, Al, and V: Application to Characterization of Al Nanoparticles Synthesized by Fs Laser Ablation. *Plasmonics* **2016**, *12*, 1813–1824. [CrossRef]
16. Scaffardi, L.B.; Tocho, J.O. Size dependence of refractive index of gold nanoparticles. *Nanotechnology* **2006**, *17*, 1309–1315. [CrossRef]
17. Schinca, D.C.; Scaffardi, L.B.; Videla, F.A.; Torchia, G.A.; Moreno, P.; Roso, L. Silver-silver oxide core-shell nanoparticles by femtosecond laser ablation: Core and shell sizing by extinction spectroscopy. *J. Phys. D Appl. Phys.* **2009**, *42*, 215102–215109. [CrossRef]
18. Santillán, J.M.J.; Videla, F.A.; Schinca, D.C.; Scaffardi, L.B.; Raap, M.B.F.V. Analysis of the structure, configuration, and sizing of Cu and Cu oxide nanoparticles generated by fs laser ablation of solid target in liquids. *J. Appl. Phys.* **2013**, *113*, 134305–134309. [CrossRef]
19. Santillán, J.M.J.; Videla, F.A.; Raap, M.B.F.V.; Muraca, D.; Scaffardi, L.B.; Schinca, D.C. Influence of size-corrected bound-electron contribution on nanometric silver dielectric function. Sizing through optical extinction spectroscopy. *J. Phys. D Appl. Phys.* **2013**, *46*, 435301. [CrossRef]
20. Kreibig, U.; Fragstein, C.V. The Limitation of Electron Mean Free Path in Small Silver Particles. *Z. Für Phys. A Hadron. Nucl.* **1969**, *224*, 1307–1323. [CrossRef]
21. Inouye, H.; Tanaka, K.; Tanahashi, I.; Hirao, K. Ultrafast dynamics of nonequilibrium electrons in a gold nanoparticle system. *Phys. Rev. B* **1998**, *57*, 11334–11340. [CrossRef]
22. Jacak, W. Lorentz Friction for Surface Plasmons in Metallic Nanospheres. *J. Phys. Chem. C* **2015**, *119*, 6749–6759. [CrossRef]
23. Jacak, W. Size-dependence of the Lorentz friction for surface plasmons in metallic nanospheres. *Opt. Express* **2015**, *23*, 4472–4481. [CrossRef] [PubMed]
24. Kluczyk, K.; Jacak, W. Damping-induced size effect in surface plasmon resonance in metallic nano-particles: Comparison of RPA microscopic model with numerical finite element simulation (COMSOL) and Mie approach. *J. Quant. Spectrosc. Radiat. Transf.* **2016**, *168*, 78–88. [CrossRef]
25. Trivedi, N.; Ashcroft, N.W. Quantum size effects in transport properties of metallic films. *Phys. Rev. B Condens. Matter* **1988**, *38*, 12298. [CrossRef] [PubMed]
26. Ruben, E.; Andrei, G.B.; Peter, N.; Javier, A. Bridging quantum and classical plasmonics with a quantum-corrected model. *Nat. Commun.* **2012**, *3*, 825.
27. ShuFen, T.; Lin, W.; Joel, K.W.Y.; Ping, B.; Michel, B.; Christian, A.N. Quantum plasmon resonances controlled by molecular tunnel junctions. *Nat. Commun.* **2014**, *343*, 1496–1499.
28. Jones, W.E.; Kliewer, K.L.; Fuchs, R. Nonlocal Theory of the Optical Properties of Thin Metallic Films. *Phys. Rev.* **1969**, *178*, 1201–1203. [CrossRef]
29. Chang, R.; Leung, P.T. Nonlocal effects on optical and molecular interactions with metallic nanoshells. *Phys. Rev. B* **2006**, *73*, 125438. [CrossRef]
30. Fuchs, R.; Claro, F. Multipolar response of small metallic spheres: Nonlocal theory. *Phys. Rev. B Condens. Matter* **1987**, *35*, 3722–3727. [CrossRef]
31. Mortensen, N.A.; Raza, S.; Wubs, M.; Sondergaard, T.; Bozhevolnyi, S.I. A generalized non-local optical response theory for plasmonic nanostructures. *Nat. Commun.* **2014**, *5*, 3809. [CrossRef] [PubMed]
32. Toscano, G.; Straubel, J.; Kwiatkowski, A.; Rockstuhl, C.; Evers, F.; Xu, H.; Mortensen, N.A.; Wubs, M. Resonance shifts and spill-out effects in self-consistent hydrodynamic nanoplasmonics. *Nat. Commun.* **2015**, *6*, 7132. [CrossRef]
33. Johnson, P.B.; Christy, R.W. Optical Constants of the Noble Metals. *Phys. Rev. B* **1972**, *6*, 4370–4379. [CrossRef]
34. Werner, W.S.M.; Glantschnig, K.; Ambrosch-Draxl, C. Optical Constants and Inelastic Electron-Scattering Data for 17 Elemental Metals. *J. Phys. Chem. Ref. Data* **2009**, *38*, 1013. [CrossRef]
35. Xiang, H.P.; Zhang, X.; Neuhauser, D.; Lu, G. Size-Dependent Plasmonic Resonances from Large-Scale Quantum Simulations. *J. Phys. Chem. Lett.* **2014**, *5*, 1163–1169. [CrossRef] [PubMed]
36. Hohenester, U.; Trügler, A. MNPBEM—A Matlab toolbox for the simulation of plasmonic nanoparticles. *Comput. Phys. Commun.* **2012**, *183*, 370–381. [CrossRef]

37. Ratnanather, J.T.; Kim, J.H.; Zhang, S.; Davis, A.M.J.; Lucas, S.K. Algorithm 935: IIPBF, a MATLAB toolbox for infinite integral of products of two Bessel functions. *Acm Trans. Math. Softw.* **2014**, *40*, 131–162. [CrossRef]
38. Boardman, A.D. *Electromagnetic Surface Modes*; Wiley: New York, NY, USA, 1982.
39. Mcmahon, J.M.; Gray, S.K.; Schatz, G.C. Calculating Nonlocal Optical Properties of Structures with Arbitrary Shape. *Phys. Rev. B* **2010**, *82*, 035423. [CrossRef]
40. Mcmahon, J.M.; Gray, S.K.; Schatz, G.C. Optical properties of nanowire dimers with a spatially nonlocal dielectric function. *Nano Lett.* **2010**, *10*, 3473–3481. [CrossRef]
41. Agarwal, G.S.; Pattanayak, D.N.; Wolf, E. Electromagnetic-Fields In Spatially Dispersive Media. *Phys. Rev. B* **1974**, *10*, 1447–1475. [CrossRef]
42. Anderegg, M.; Feuerbacher, B.; Fitton, B. Optically Excited Longitudinal Plasmons in Potassium. *Phys. Rev. Lett.* **1971**, *27*, 1565–1568. [CrossRef]
43. Coronado, E.A.; Schatz, G.C. Surface plasmon broadening for arbitrary shape nanoparticles: A geometrical probability approach. *J. Chem. Phys.* **2003**, *119*, 3926–3934. [CrossRef]
44. Pinchuk, A.; Kreibig, U.; Hilger, A. Optical properties of metallic nanoparticles: Influence of interface effects and interband transitions. *Surf. Sci.* **2004**, *557*, 269–280. [CrossRef]
45. Alvarez, M.M.; Khoury, J.I.; Schaaff, T.G.; Shafigullin, M.N.; Vezmar, I.; Whetten, R.L. Optical Absorption Spectra of nanocrystal gold molecules. *J. Phys. Chem. B* **1997**, *101*, 3706–3712. [CrossRef]
46. Rosei, R.; Antonangeli, F.; Grassano, U.M. d bands position and width in gold from very low temperature thermomodulation measurements. *Surf. Sci.* **1973**, *37*, 689–699. [CrossRef]
47. Sha, W.E.I.; Liu, A.Y.; Weng, C.C. Dissipative Quantum Electromagnetics. *IEEE J. Multiscale Multiphy. Comput. Tech.* **2018**, *3*, 198–213. [CrossRef]

*applied sciences*

MDPI

*Article*

# Evidence-Theory-Based Robust Optimization and Its Application in Micro-Electromechanical Systems

**Zhiliang Huang [1,2], Jiaqi Xu [1], Tongguang Yang [1], Fangyi Li [3,*] and Shuguang Deng [1]**

1 School of Mechanical and Electrical Engineering, Hunan City University, Yiyang 413002, China; 13787181710@163.com (Z.H.); 18552122421@163.com (J.X.); yangtongguang1@163.com (T.Y.); shuguangdeng@163.com (S.D.)
2 College of Mechanical and Vehicle Engineering, Hunan University, Changsha 40082, China
3 School of Vehicle and Mechanical Engineering, Changsha University of Science and Technology, Changsha 410076, China
* Correspondence: fangyi.li@csust.edu.cn

Received: 21 February 2019; Accepted: 1 April 2019; Published: 7 April 2019

**Featured Application: This paper develops an evidence-theory-based robustness optimization (EBRO) method, which aims to provide a potential computational tool for engineering problems with epistemic uncertainty. This method is especially suitable for robust designing of micro-electromechanical systems (MEMS). On one hand, unlike traditional engineering structural problems, the design of MEMS usually involves micro structure, novel materials, and extreme operating conditions, where multi-source uncertainties inevitably exist. Evidence theory is well suited to deal with such uncertainties. On the other hand, high performance and insensitivity to uncertainties are the fundamental requirements for MEMS design. The robust optimization can improve performance by minimizing the effects of uncertainties without eliminating these causes.**

**Abstract:** The conventional engineering robustness optimization approach considering uncertainties is generally based on a probabilistic model. However, a probabilistic model faces obstacles when handling problems with epistemic uncertainty. This paper presents an evidence-theory-based robustness optimization (EBRO) model and a corresponding algorithm, which provide a potential computational tool for engineering problems with multi-source uncertainty. An EBRO model with the twin objectives of performance and robustness is formulated by introducing the performance threshold. After providing multiple target belief measures (Bel), the original model is transformed into a series of sub-problems, which are solved by the proposed iterative strategy driving the robustness analysis and the deterministic optimization alternately. The proposed method is applied to three problems of micro-electromechanical systems (MEMS), including a micro-force sensor, an image sensor, and a capacitive accelerometer. In the applications, finite element simulation models and surrogate models are both given. Numerical results show that the proposed method has good engineering practicality due to comprehensive performance in terms of efficiency, accuracy, and convergence.

**Keywords:** epistemic uncertainty; evidence theory; robust optimization; sensor design

## 1. Introduction

In practical engineering problems, various uncertainties exist in terms of the operating environment, manufacturing process, material properties, etc. Under the combined action of these uncertainties, the performance of engineering structures or products may fluctuate greatly. Robust optimization [1,2] is a methodology, and its fundamental principle is to improve the performance of a product by minimizing the effects of uncertainties without eliminating these causes. The concept of robustness optimization has long been embedded in engineering design. In recent years, thanks to the rapid development of computer technology, it has been widely applied to many engineering fields, such as electronic [3], vehicle [4], aerospace [5], and civil engineering [6]. The core of robustness optimization lies in understanding, measuring, and controlling the uncertainty in the product design process. In mechanical engineering disciplines, uncertainty is usually differentiated into objective and subjective from an epistemological perspective [7]. The former, also called aleatory uncertainty, comes from an inherently irreducible physical nature, e.g., material properties (elasticity modulus, thermal conductivity, expansion coefficient) and operating conditions (temperature, humidity, wind load). A probabilistic model [8–10] is an appropriate way to describe such uncertain parameters, provided that sufficient samples are obtained for the construction of accurate random distribution. Conventional robustness optimization methods [11–13] are based on probabilistic models, in which the statistical moments (e.g., mean, variance) are employed to formulate the robustness function for the performance assessment under uncertainties. On the other hand, designers may lack knowledge about the issues of concern in practice, which leads to subjective uncertainty, also known as epistemic uncertainty. The uncertainty is caused by cognitive limitation or a lack of information, which could be reduced theoretically as effort is increased. At present, the methods of dealing with epistemic uncertainty mainly include possibility theory [14,15], the fuzzy set [16,17], convex model [18,19], and evidence theory [20,21]. Among them, evidence theory is an extension of probability theory, which can properly model the information of incompleteness, uncertainty, unreliability and even conflict [22]. When evidence theory treats a general structural problem, all possible values of an uncertain variable are assigned to several sub-intervals, and the corresponding probability is assigned to each sub-interval according to existing statistics and expert experience. After synthesizing the probability of all the sub-intervals, the belief measure and plausibility measure are obtained, which constitute the confidence interval of the proposition and show that the structural performance satisfies a given requirement. Compared with other uncertainty analysis theories, evidence theory may be more general. For example, when the sub-interval of each uncertain variable is infinitely small, evidence theory is equivalent to probability theory; when the sub-interval is unique, it is equivalent to convex model theory; when no conflict occurs to the information from different sources, it is equivalent to possibility theory [23].

In the past decade, some progress has been made in evidence-theory-based robust optimization (EBRO). For instance, Vasile [24] employed evidence theory to model the uncertainties of spacecraft subsystems and trajectory parameters in the robust design of space trajectory and presented a hybrid co-evolutionary algorithm to obtain the optimal results. For the preliminary design of a space mission, Croisard et al. [25] formulated the robust optimization model using evidence theory and proposed three practical solving technologies. Their features of efficiency and accuracy were discussed through the application of a space mission. Zuiani et al. [26] presented a multi-objective robust optimization approach for the deflection action design of near-Earth objects, and the uncertainties involved in the orbital and system were qualified by evidence theory. A deflection design application of a spacecraft swarm with Apophis verified the effectiveness of this approach. Hou et al. [27] introduced evidence-theory-based robust optimization (EBRO) into multidisciplinary aerospace design, and the strategy of an artificial neural network was used to establish surrogate models for the balance of efficiency and accuracy during the optimization. This method was applied to two preliminary designs of the micro entry probe and orbital debris removal system. The above studies employed evidence theory to measure the epistemic uncertainties involved in engineering design, and expanded robustness optimization into the design of complex systems. However, the studies of EBRO are still in a preliminary stage. The existing research has mainly aimed at the preliminary design of engineering systems. Most of them have been simplified and assumed to a great extent. In other words, the performance functions are based on surrogate models and even empirical formulas. So far, EBRO applications in actual product design, with a time-consuming simulation model being created for performance function, are actually quite few. After all, computational cost is a major technical bottleneck limiting EBRO applications. First, evidence theory describes uncertainty through a series of discontinuous sets, rather than a continuous function similar to a probability density function. This usually leads to a combination explosion in a multidimensional robustness analysis, and finally results in a heavy computational burden. Secondly, EBRO is essentially a nested optimization problem with performance optimization in the outer layer and robustness analysis in the inner layer. The direct solving strategy means a large number of robustness evaluations using evidence theory. As a result, the issue of EBRO efficiency is further exacerbated. Therefore, there is a great engineering significance in developing an efficient EBRO method in view of actual product design problems.

In this paper, a general EBRO model and an efficient algorithm are proposed, which provide a computational tool for robust product optimization with epistemic uncertainty. The proposed method is applied to three design problems of MEMS, in which its engineering practicability is discussed. The remainder of this paper is organized as follows. Section 1 briefly introduces the basic concepts and principles of robustness analysis using evidence theory. The EBRO model is formulated Section 2. The corresponding algorithm is proposed in Section 3. In Section 4, this method is validated through the three applications of MEMS—a micro-force sensor, a low-noise image sensor and a capacitive accelerometer. Conclusions are drawn in Section 5.

## 2. Robustness Analysis Using Evidence Theory

Consider that uncertainty problem is given as $f(\mathbf{Z})$, where $\mathbf{Z}$ represents the $n_Z$-dimensional uncertain vector, $f$ is the performance function which is uncertain due to $\mathbf{Z}$. Conventional methods [16–18] of robust optimization employ probability theory to deal with the uncertainties. The typical strategy is to consider the uncertain parameters of a problem as random variables, and thereby the performance value is also a random variable. The mean and variance are used to formulate the robustness model. In practical engineering, it is sometimes hard to construct accurate probability models due to the limited information. Thus, evidence theory [20,21] is adopted to model the robustness. In evidence theory, the frame of discernment (FD) needs to be established first, which contains several independent basic propositions. It is similar to the sample space of a random parameter in probability theory. Here, $2^\Theta$ denotes the power set of the FD (namely $\Theta$), and $2^\Theta$ consists of all possible propositions contained in $\Theta$. For example, for a FD with the two basic propositions of $\Theta_1$ and $\Theta_2$, the corresponding power set is $2^\Theta = \{\varnothing, \{\Theta_1\}, \{\Theta_2\}, \{\Theta_1, \Theta_2\}\}$. Evidence theory adopts a basic probability assignment (BPA) to measure the confidence level of each proposition. For a certain proposition $A$, the BPA is a mapping function that satisfies the following axioms:

$$\begin{array}{l} 0 \leq m(A) \leq 1,\ \forall A \in 2^\Theta \\ m(\Phi) = 0 \\ \sum\limits_{A \in 2^\Theta} m(A) = 1 \end{array} \tag{1}$$

where if $m(A) \geq 0$, $A$ is called a focal element of $m$. The BPA of $m(A)$ denotes the extent to which the evidence supports Proposition A. When the information comes from multiple sources, $m(A)$ can be obtained by evidence combination rules [28]. Evidence theory uses an interval consisting of the belief measure (*Bel*) and the plausibility measure (*Pl*) to describe the true extent of the proposition. The two measures are defined as:

$$\begin{array}{l} Bel(A) = \sum\limits_{C \subseteq A} m(C) \\ Pl(A) = \sum\limits_{C \cap A \neq \Phi} m(C) \end{array} \tag{2}$$

As can be seen from Equation (2), $Bel(A)$ is the summary of all the BPA that totally support Proposition $A$, while $Pl(A)$ is the summary of the BPA that support Proposition $A$ totally or partially.

A two-dimensional design problem is taken as the example to illustrate the process of robustness analysis using evidence theory. The performance function contains two uncertain parameters $(a, b)$, which are both considered as evidence variables. The FDs of $a$, $b$ are the two closed intervals, i.e., $A = \left[A^L, A^R\right]$ and $B = \left[B^L, B^R\right]$. $A$ contains $n_A$ number of focal elements, and the subinterval of $A_i = \left[B_i^L, B_i^R\right]$ represents the $i$-th focal element of $A$. The definitions of $n_B$ and $B_j$ are similar. Thus, a Cartesian product can be constructed:

$$D = A \times B = \left\{D_k = \left[A_i, B_j\right], A_i \in A, B_j \in B\right\} \tag{3}$$

where $D_k$ is the $k$-th focal element of $D$, and the total number of focal elements is $n_A \times n_B$. For ease of presentation, assuming that $a$, $b$ are independent, a two-dimensional joint BPA is obtained:

$$m(D_k) = m(A_i) \cdot m(B_j) \tag{4}$$

More general problems with parametric correlation can be handled using the mathematical tool of copula functions [29].

As analyzed above, the performance function of $f$ is uncertain. The performance threshold of $v$ is given to evaluate its robustness. Given that the design objective is to minimize the value of $f$, the higher the trueness of Proposition $f \le v$, the higher the robustness of $f$ relative to $v$. Proposition $f \le v$ is defined as the feasible domain:

$$F = \{f : f(a,b) \le v\}, (a,b) \in D_k, D_k = \left[A_i, B_j\right] \subset D \tag{5}$$

Substituting $A$, $C$ with $F$, $D_k$ in Equation (2), the belief measure and plausibility measure of Proposition $f \le v$ are expressed as follows:

$$\begin{aligned} Bel(F) &= \sum_{D_k \subseteq F} m(D_k) \\ Pl(F) &= \sum_{D_k \cap F \neq \Phi} m(D_k) \end{aligned} \tag{6}$$

In evidence theory, the probabilistic interval composed by the two measures can describe the trueness of $f \le v$, written as $R(F) \in [Bel(F), Pl(F)]$. The accumulation of $Bel$, $Pl$ needs to determine the positional relationship between each focal element and the $F$ domain. As a result, the performance function extrema of each focal element must be searched. For this example, the $n_A \times n_B$ pairs of extremum problems are established as:

$$\left.\begin{aligned} f_k^{\min} &= \min_{(a,b) \in D_k} f(a,b) \\ f_k^{\max} &= \max_{(a,b) \in D_k} f(a,b) \end{aligned}\right\} k = 1, 2, \ldots, n_A \times n_B \tag{7}$$

where $f_k^{\min}$, $f_k^{\max}$ are the minimum and maximum of the $k$-th focal element. The vertex method [30] can efficiently solve the problems in Equation (7) one by one. If $f_k^{\max} \le v$, $D_k \subseteq F$, and $m(D_k)$ is simultaneously accounted into $Bel(F)$ and $Pl(F)$; If $f_k^{\min} < v$, $D_k \cap F \neq \Phi$, and $m(D_k)$ is only accounted into $Pl(G)$. After calculating the extrema for all focal elements, the $Bel$ and $Pl$ can be totaled.

## 3. Formulation of the EBRO Model

As mentioned above, evidence theory uses a pair of probabilistic values $[Bel, Pl]$ to measure the robustness of the performance value related to the given threshold. However, engineers generally tend to adopt conservative strategies to deal with uncertainties in the product design process. Thus, the robustness objective of EBRO can be established as $\max Bel(f \le v)$. Meanwhile, in order to improve product performance, the performance threshold is minimized. The EBRO model is formulated as a double-objective optimization problem:

$$\begin{aligned} &\min v,\ \max Bel(f(\boldsymbol{d},\boldsymbol{X},\boldsymbol{P}) \le v) \\ &\text{s.t. } \boldsymbol{d}^l \le \boldsymbol{d} \le \boldsymbol{d}^u,\ \overline{\boldsymbol{X}}^l \le \boldsymbol{X} \le \overline{\boldsymbol{X}}^u \end{aligned} \tag{8}$$

where $\boldsymbol{d}$ is the $n_d$-dimensional deterministic design vector; $\boldsymbol{X}$ is the $n_X$-dimensional uncertain design vector; $\boldsymbol{P}$ is the $n_P$-dimensional uncertain parameter vector; the superscripts of $l$, $u$ represent the value range of a design variable; and $\overline{\boldsymbol{X}}$ represents the nominal value of $\boldsymbol{X}$. Note that the threshold of $v$ is usually difficult to give a fixed value to, while it should be treated as a deterministic design variable.

The proposed model is an improvement on the existing model [24] because it can handle more types of uncertainty, such as the perturbations of design variables resulting from production tolerances, and the variations of parameters due to changing operating conditions. As for the solving process, the EBDO involves the nested optimization of the double-objective optimization in the outer layer and the robustness assessment in the inner layer. Due to the discreteness introduced by the evidence variables, each of the robustness analyses need to calculate the performance extrema of all focal elements. Essentially, extremum evaluation is an optimization problem involving the performance function based on time-consuming simulation models, and therefore the robustness analysis bears a high computational cost. More seriously, the double-objective optimization in the outer layer requires a large number of robustness evaluations in the inner layer. Eventually, the EBDO solving becomes extremely inefficient.

## 4. The Proposed Algorithm

To improve efficiency, this paper proposes a decoupling algorithm of EBRO, and its basic idea is to convert the nested optimization into the sequence iteration process. Firstly, the original problem is decomposed into a series of sub-problems. Secondly, the uncertainty analysis and the deterministic optimization are driven alternately until convergence. The framework of the proposed method is detailed below.

### *4.1. Decomposition into Sub-Problems*

Robust optimization is essentially a multi-objective problem that increases product performance at the expense of its robustness. Therefore, robust optimization generally does not have a unique solution, but a set of solutions called the Pareto optimal set [2]. It is a family of solutions that is optimal in the sense that no improvement can be achieved in any objective without degradation in others for a multi-objective problem. The Pareto-optimal solutions can be obtained by solving appropriately formulated single objective optimization problems on a one-at-a-time basis. At present, a number of multi-objective genetic algorithms have been suggested. The primary reason for this is their ability to find multiple Pareto-optimal solutions in parallel. From the viewpoint of mathematical optimization, genetic algorithms are a kind of suitable method for solving a general multi-objective optimization. However, the efficiency of a genetic algorithm is usually much lower than the gradient-based optimization algorithms, which has become the main technical bottleneck limiting its practical application [31,32]. Although a priori information is not required when using genetic algorithms, most designers have some engineering experience in practice. Therefore, for the specific problem shown in Equation(8), the robustness objective of $Bel(f(\boldsymbol{d},\boldsymbol{X},\boldsymbol{P}) \le v)$ is often handled as a reliability constraint [2,33]. In this paper, the EBRO problem is transformed into a series of sub-problems under the given target belief measures:

$$\left.\begin{aligned} &\min v \\ &\text{s.t. } Bel(f(\boldsymbol{d},\boldsymbol{X},\boldsymbol{P}) \le v) \ge Bel_j^T \\ &\quad \boldsymbol{d}^l \le \boldsymbol{d} \le \boldsymbol{d}^u,\ \overline{\boldsymbol{X}}^l \le \boldsymbol{X} \le \overline{\boldsymbol{X}}^u \end{aligned}\right\} j = 1,2,\ldots,n_T \tag{9}$$

where $Bel_j^T$ represents the $j$-th target belief measure; and $Bel(f \leq v) \geq Bel_j^T$ is the reliability constraint derived from the robust objective. In many cases, the designer may focus on the performance values under some given conditions based on the experience or quality standard. This condition is usually a certain probability of $f \leq v$, namely $Bel_j^T$.

### 4.2. Iteration Framework

Theoretically, the EBDO problems in Equation (9) can be solved by existing methods [34]. However, the resulting computational burden will be extremely heavy. To address this issue, a novel iteration framework is developed, in which the uncertainty analysis and design optimization alternate until convergence.

In the $k$-th iteration, each optimization problem in Equation (9) requires the performance of an uncertainty analysis at the previous design point:

$$Bel\left(f(\mathbf{Z}) \leq v_j^{(k-1)}\right), \mathbf{Z} = (\mathbf{X}, \mathbf{P}), j = 1, 2, \ldots, n_T \tag{10}$$

This mainly consists of two steps, illustrated by the example in Figure 1. Step 1 is to search for the most probable focal element (MPFE) along the limit-state boundary of $f(\mathbf{Z}) \leq v_j^{(k-1)}$. The MPFE [35] is similar to the most probable point (MPP) in probability theory, which is the point with the most probability density on the limit-state boundary. Compared to other points on the boundary, the minimal error of reliability analysis can be achieved by establishing the linear approximation for the performance function at the MPP [36]. Similarly, the MPFE contains the maximal BPA among the focal elements that are crossed by the limit-state boundary. The searching process of MPFE is formulated as:

$$\left.\begin{array}{l} \max\limits_{D_k} m(D_k) \\ \text{s.t. } f(\mathbf{Z}) = v_j^{(k-1)} \end{array}\right\} j = 1, 2, \ldots, n_T \tag{11}$$

where $m(D_k)$ represents the BPA of the focal element where the $\mathbf{Z}$ point is located. Note that there is a difference between $v_j^{(k-1)}, j = 1, 2, \ldots, n_T$ at each iteration step due to the minor difference of $Bel_j^T$. Consequently, different MPFEs may be obtained for Equation (11). However, the difference between the MPFEs is minor relative to the entire design domain. To ensure efficiency, the unique MPFE is investigated at each iteration. Equation (11) can be rewritten as:

$$\begin{array}{l} \max\limits_{D_k} m(D_k) \\ \text{s.t. } f(\mathbf{Z}) = v^{(k-1)} \end{array} \tag{12}$$

where $v^{(k-1)}$ represents the performance threshold that has not yet converged.

Step 2 is to establish linear approximation for the performance function at the central point $\left(\overset{-M}{\mathbf{Z}}\right)$ of MPFE:

$$L^{(k)}(\mathbf{Z}) = f\left(\mathbf{Z}^{M(k)}\right) + \left(\mathbf{Z} - \mathbf{Z}^{M(k)}\right)^{\mathrm{T}} \cdot \nabla f\left(\mathbf{Z}^{M(k)}\right) \tag{13}$$

The $L$-function is used to replace the $f$-function to calculate $Bel$, and thereby the optimization processes in Equation (7) no longer requires the calculation of any performance function.

The efficient calculation of $Bel$ has been achieved in the iterative process, but the overall process of EBDO still requires dozens or even hundreds of $Bel$ evaluations due to the nested optimization. To eliminate the nested optimization, a decoupling strategy is proposed similar to that in the probabilistic method [37]. At each iteration step, the reliability constraint is transformed into a deterministic constraint by constructing the shifting vector of $\mathbf{S}_j^{(k)}$; and then a deterministic optimization is updated and solved to obtain the current solution. In the $k$-th iteration, the deterministic optimization can be written as:

$$\left.\begin{array}{l} \min v \\ \text{s.t. } f\left(\boldsymbol{d}, \bar{\mathbf{Z}} - \mathbf{S}_j^{(k)}\right) \le v \\ \quad \boldsymbol{d}^l \le \boldsymbol{d} \le \boldsymbol{d}^u,\ \overline{\mathbf{X}}^l \le \mathbf{X} \le \overline{\mathbf{X}}^u \end{array}\right\} j = 1, 2, \ldots, n_T \tag{14}$$

The shifting vector determines the deviation between the original reliability boundary and the deterministic boundary at the $k$-th iteration step. For the $j$-th problem in Equation (14), the formulation of the shifting vector is explained as in Figure 2. For convenience of presentation, the constraint contains only two evidence variables $\mathbf{Z} = (a, b)$. $F$ represents the domain of $f \le v_j^{(k-1)}$. $\bar{\mathbf{Z}}_j^{(k-1)}$ is the previous design point, which is based on the previous equivalent boundary of $f\left(\mathbf{Z} - \mathbf{S}_j^{(k-1)}\right) = v_j^{(k-1)}$. The rectangular domain represents the FD at the previous design point. $F$ represents the domain of $f \le v_j^{(k-1)}$. If the FD is entirely in the $F$ domain, $Bel\left(f \le v_j^{(k-1)}\right) = 100\%$. In Figure 2, the FD of $\bar{\mathbf{Z}}^{(k-1)}$ is partially in the $F$ domain, and $Bel\left(f \le v_j^{(k-1)}\right)$ is still less than $Bel_j^T$. To satisfy $Bel\left(f(\mathbf{Z}) \ge v_j^{(k-1)}\right) \ge Bel_j^T$, $\bar{\mathbf{Z}}^{(k-1)}$ needs to move further into the $F$ domain. Therefore, the equivalent boundary needs to move further toward the $F$ domain. The updated equivalent boundary is constructed as follows:

$$f\left(\mathbf{Z} - \mathbf{S}_j^{(k)}\right) = v_j^{(k-1)},\ \mathbf{S}_j^{(k)} = \mathbf{S}_j^{(k-1)} + \Delta\mathbf{S}_j^{(k)} \tag{15}$$

where $\Delta\mathbf{S}_j^{(k)}$ denotes the increment of the previous shifting vector. The principle for calculating $\Delta\mathbf{S}_j^{(k)}$ is set as $Bel\left(f(\mathbf{Z}) \ge v_j^{(k-1)}\right) \ge Bel_j^T$ and is just satisfied. Thus, the mathematical model of $\Delta\mathbf{S}_j^{(k)}$ is created as:

$$\begin{array}{c} \min_{\boldsymbol{s}} \| \boldsymbol{s} \| \\ \text{s.t. } Bel\left(f(\mathbf{Z} + \boldsymbol{s}) \le v_j^{(k-1)}\right) = Bel_j^T \end{array} \tag{16}$$

Equation (16) can be solved by multivariable optimization methods [38]. To further improve efficiency, the $f$-function is replaced by the $L$-function formulated in Equation (13).

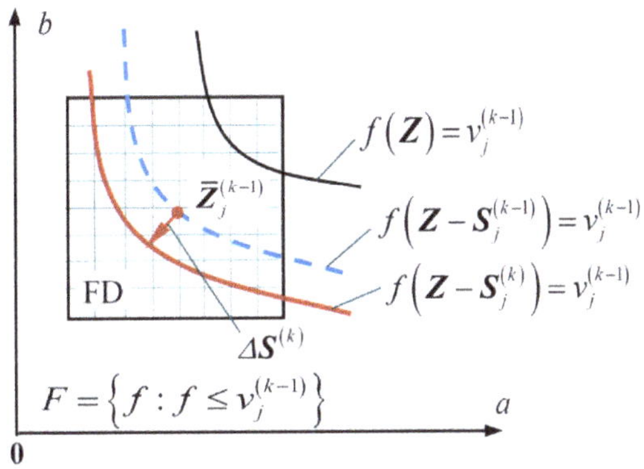


**Figure 1.** Uncertainty analysis for the performance function. FD: frame of discernment; MPFE: most probable focal element.

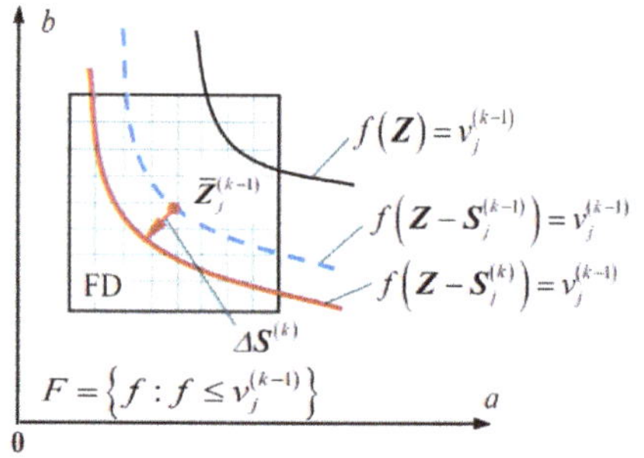


**Figure 2.** Formulation of the shifting vector. FD: frame of discernment.

Uncertain analysis and design optimization are carried out alternatively until they meet the following convergence criteria:

$$\left.\begin{array}{l} Bel_j^{(k)} \geq Bel_j^T \\ \left|\frac{v_j^{(k)}-v_j^{(k-1)}}{v_j^{(k)}}\right| \leq \varepsilon_r \end{array}\right\} j=1,2,\ldots,n_T \tag{17}$$

where $\varepsilon_r$ is the minimal error limit. The solutions of $\left(d_j^*, \overline{X}_j^*\right)$, $j = 1,2,\ldots,n_T$ form the final optimal set.

The flowchart of the EBRO algorithm is summarized as Figure 3.

## 5. Application Discussion

In the previous sections, an EBDO method is developed for engineering problems with epistemic uncertainty. This method is especially suitable for the robust design of micro-electromechanical systems (MEMS). On one hand, unlike traditional engineering structural problems, the design of MEMS usually involves micro structure, novel materials, and extreme operating conditions, where epistemic uncertainties inevitably exist. Evidence theory is well suited to deal with such uncertainties. On the other hand, high performance and insensitivity to uncertainties are the fundamental requirements for MEMS design. Over the past two decades, robust optimization for MEMS has gradually attracted the attention of both academics and engineering practice [39–41]. In this section, this method is applied to three applications of MEMS: a micro-force sensor, a low-noise image sensor, and a capacitive accelerometer. The features of the proposed approach are investigated in terms of efficiency and accuracy. Performance function evaluations are accounted to indicate efficiency, and the reference solution is compared to verify accuracy. The reference solution is obtained by the double-loop method, where sequential quadratic programming [38] is employed for performance optimization, and a Monte-Carlo simulation [42] is used for robust assessment.

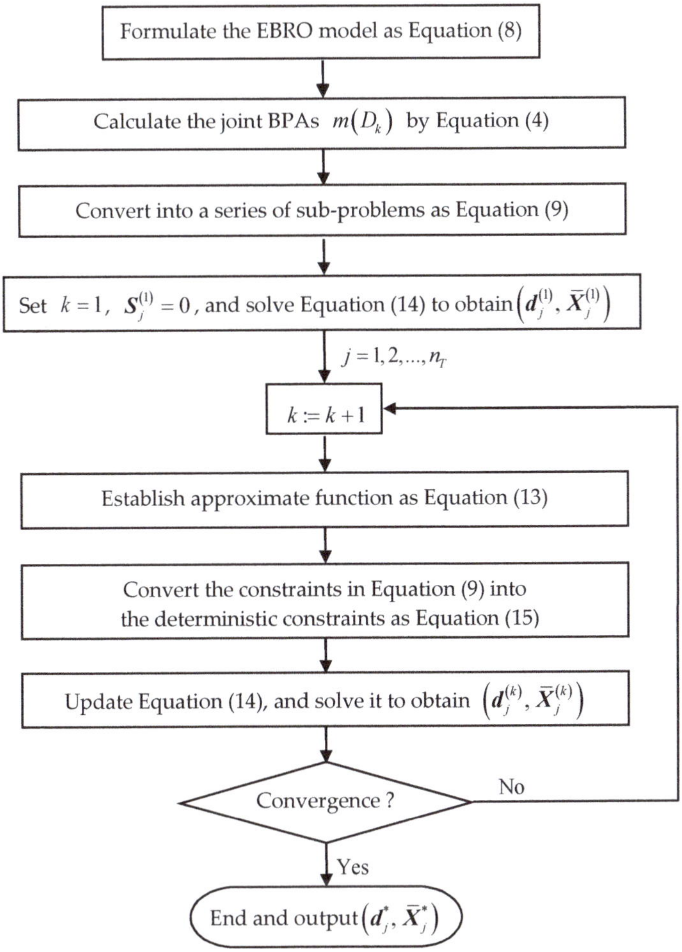


**Figure 3.** The flowchart of the proposed method. EBRO: evidence-theory-based robust optimization; BPA: basic probability assignment.

### 5.1. A Micro-Force Sensor

A piezoelectric micro-force sensor [43] has several advantages, including a reliable structure, fast response, and simple driving circuits. It has been extensively applied in the fields of precision positioning, ultrasonic devices, micro-force measurement, etc. Given that uncertainties are inevitable in structural sizes and material parameters, robust optimization is essential to ensure the performance of the sensor.

As shown in Figure 4, the core part of the micro-force sensor is a piezoelectric cantilever beam, which consists of a piezoelectric film, a silicon-based layer, and two electrodes. The force at the free end causes bending deformation on the beam, which drives the piezoelectric film to output polarization charges through the piezoelectric effect. The charge is transmitted to the circuit by the electrodes and converted into a voltage signal. According to the theoretical model proposed by Smits et al. [43], this voltage can be formulated as:

$$U=\frac{3\cdot d_{31}^{P}\cdot S_{11}^{Si}\cdot S_{11}^{P}\cdot h\cdot h^{P}\cdot\left(h+h^{P}\right)\cdot L\cdot F}{K\cdot \varepsilon_{33}^{P}\cdot w} \tag{18}$$

where

$$K = 4 \cdot S_{11}^{Si} \cdot S_{11}^{P} \cdot h \cdot \left(h^P\right)^3 + 4 \cdot S_{11}^{Si} \cdot S_{11}^{P} \cdot h^3 \cdot h^P + \left(S_{11}^{P}\right)^2 \cdot h^4 + \left(S_{11}^{Si}\right)^2 \cdot \left(h^P\right)^4 + 4 \cdot S_{11}^{Si} \cdot S_{11}^{P} \cdot \left(S_{11}^{Si}\right)^2 \cdot \left(h^P\right)^2 \tag{19}$$

where $F$ is the concentration force; $L$, $w$ represent the length and width of the beam; $h$, $h^P$ denote the thickness of the silicon-base layer and piezoelectric film; $S_{11}^{Si}$, $S_{11}^{P}$ are the compliance coefficient of the silicon-based layer and piezoelectric film; and $d_{31}^{P}$, $\varepsilon_{33}^{P}$ is the piezoelectric coefficient and dielectric constant of the piezoelectric film. The constants in Equation (18) include $h^P = 5 \times 10^{-4}$ mm, $S_{11}^{P} = 18.97 \times 10^{-12} \mathrm{m}^2/\mathrm{N}$, and $S_{11}^{Si} = 7.70 \times 10^{-12} \mathrm{m}^2/\mathrm{N}$, The structural sizes of $L$, $w$, $h$ and the material parameter of $d_{31}^{P}$, $\varepsilon_{33}^{P}$ are viewed as evidence variables. The marginal BPAs of the variables are shown in Figure 5, and the nominal values of $d_{31}^{P}$, $\varepsilon_{33}^{P}$ are, respectively, 1.8 C/N, 1.6 F/m.

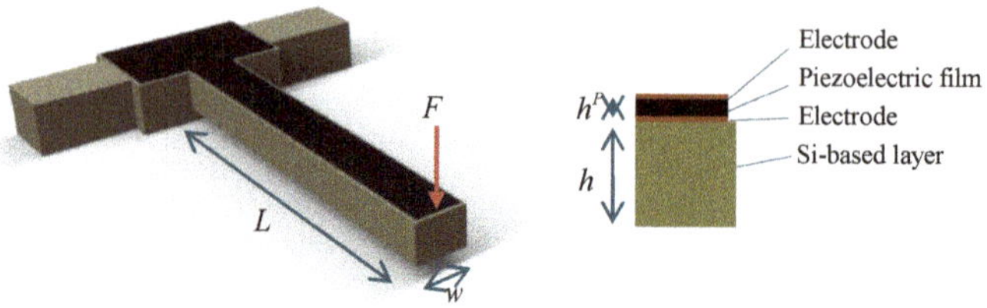


**Figure 4.** A piezoelectric cantilever beam.

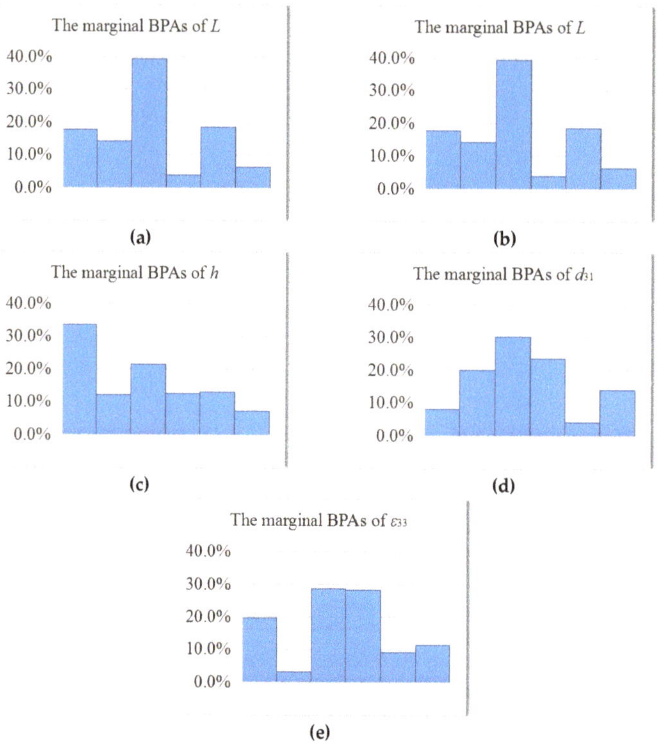


**Figure 5.** Marginal BPAs of variables in the micro-force sensor problem. BPA: basic probability assignment; FD: frame of discernment.

In engineering, the greater the output voltage, the higher the theoretical accuracy of the sensor. Thus, $U$ is regarded as the objective function. The design variables are $L$, $w$ and $h$. The constraints of shape, stiffness and strength are considered, which are expressed as $\eta \geq 0.83$, $\delta \leq 2.5\,\mu\text{m}$, and $\sigma \leq 32.0\,\text{MPa}$, where $\eta$ is the ratio of $w$ to $h$, $\delta$ denotes the displacement at the free end of the beam, and $\sigma$ denotes the maximum stress of the beam. $\delta$ and $\sigma$ can be written as [43].

$$\begin{aligned} \sigma &= \frac{6\cdot F\cdot L\cdot S_{11}^{Si}\cdot\left(S_{11}^{P}\cdot h+S_{11}^{Si}\cdot h^{P}\right)\cdot\left(h+h^{P}\right)}{K\cdot w} \\ \delta &= \frac{4\cdot F\cdot L^{3}\cdot S_{11}^{Si}\cdot S_{11}^{P}\cdot\left(S_{11}^{P}\cdot h+S_{11}^{Si}\cdot h^{P}\right)}{K\cdot w} \end{aligned} \tag{20}$$

Due to uncertainties in the structure, $\eta$, $\delta$ and $\sigma$ are also uncertain. Theoretically, the three constraints should be modeled as reliability constraints. To focus on the topic of robust optimization, the constraints are considered as deterministic in this example. That is, the nominal values of the uncertain variables are used to calculate $\eta$, $\delta$ and $\sigma$. In summary, the EBRO problem is formulated as follows:

$$\begin{aligned} &\max U_0,\ \max Bel(U(\mathbf{X},\mathbf{P}) \geq U_0) \\ \text{s.t. } &\tfrac{w}{h} \leq 0.83,\ \delta\left(\overline{\mathbf{X}}\right) \leq 2.5\mu\text{m},\ \sigma\left(\overline{\mathbf{X}}\right) \leq 32.0\text{MPa} \\ &0.40\text{mm} \leq \overline{L} \leq 1.20\text{mm}, 0.06\text{mm} \leq \overline{w} \leq 0.10\text{mm}, 0.04\text{mm} \leq \overline{h} \leq 0.10\text{mm} \end{aligned} \tag{21}$$

where $\mathbf{X} = (L,\ w,\ h)$, $\mathbf{P} = (d_{31}^{P}, \varepsilon_{33}^{P})$; $U_0$ represents the performance threshold, which is set as the deterministic design variable.

The steps to solve this problem using the proposed method are detailed below. Firstly, according to the marginal BPAs of the five variables in Figure 5, the joint BPAs of the focal elements ($8^5$ = 32768) are calculated by Equation (4). Secondly, Equation (21) is converted into a series of sub-problems, which are expressed as:

$$\begin{aligned} &\left.\begin{aligned} &\max U_0 \\ &\text{s.t. } Bel(U(\mathbf{X},\mathbf{P}) \geq U_0) \geq Bel_j^T \end{aligned}\right\} j = 1,2,\ldots,5 \\ &\tfrac{w}{h} \leq 0.83,\quad \delta\left(\overline{\mathbf{X}}\right) \leq 2.5\mu\text{m},\quad \sigma\left(\overline{\mathbf{X}}\right) \leq 32.0\text{MPa} \\ &0.40\text{mm} \leq \overline{L} \leq 1.20\text{mm}, 0.06\text{mm} \leq \overline{w} \leq 0.10\text{mm}, 0.04\text{mm} \leq \overline{h} \leq 0.10\text{mm} \\ &Bel_j^T = (80\%,\ 85\%,\ 90\%,\ 95\%,\ 99.9\%) \end{aligned} \tag{22}$$

where $Bel_j^T$ represent a series of target $Bel$ for the proposition of $U \geq U_0$, which are given by the designer according to engineering experience or quality standards. Thirdly, the iteration starts from the initial point of $\left(\overline{L}^{(0)}, \overline{w}^{(0)}, \overline{h}^{(0)}, U_0^{(0)}\right) = (0.60\,\text{mm}, 0.08\,\text{mm}, 0.06\,\text{mm}, 35.6\,\text{mV})$, where $L^{(0)}, w^{(0)}, h^{(0)}$ are selected by the designer and $U_0^{(0)}$ is calculated by Equation (18). At each iteration step, the approximate function of $U$ is established as Equation (13), and then 10 numbers of $\Delta S_j^{(k)}$ are obtained through Equation (16). Correspondingly, the 10 optimization problems as Equation (14) are updated. By solving them, the optimal set in the current iteration is obtained. After four iteration steps, the optimal set is converged as listed in Table 1. The results show that the performance threshold decreases gradually with increase in *Bel*. In engineering, a designer can intuitively select the optimal design option from the optimal set by balancing the product performance and robustness. In term of accuracy, the solutions of the proposed method are very close to the corresponding reference solutions, and the maximal error is only 2.5% under the condition of $Bel_j^T = 95\%$. In efficiency, the proposed method calculates performance function only 248 times, and the computational cost is much less than that of evolutionary algorithms [31]. From a mathematical point of view, it is unfair to compare the efficiency of the proposed method with the evolutionary algorithms. From the view of engineering practicality, however, the solutions of the proposed method may help the designers create a relatively clear picture of the problem with high efficiency and acceptable accuracy.

**Table 1.** Optimal set of the micro-force sensor problem.

| Results | | $Bel_1^T = 80\%$ | $Bel_2^T = 85\%$ | $Bel_3^T = 90\%$ | $Bel_4^T = 95\%$ | $Bel_5^T = 99.9\%$ |
|---|---|---|---|---|---|---|
| Proposed method | $X_j^*$ (mm) | 0.937, 0.086, 0.072 | 0.937, 0.086, 0.072 | 0.937, 0.086, 0.072 | 0.937, 0.086, 0.072 | 0.937, 0.086, 0.072 |
| | $v_j^*, Bel_j^*$ | 22.1 mV, 80.4% | 20.0 mV, 85.3% | 18.2 mV, 90.6% | 15.7 mV, 95.5% | 9.6 mV, 99.9% |
| Reference solution | $v_j^r, Bel_j^r$ | 22.4 mV, 80.0% | 20.0 mV, 85.3% | 18.5 mV, 90.0% | 16.1 mV, 95.0% | 9.6 mV, 99.9% |

### 5.2. An Ultra-Low-Noise Image Sensor

Recently, a type of ultra-low-noise image sensor [44] was developed for applications requiring high-quality imaging under extremely low light conditions. Such a type of sensor is ideally suited for a variety of low light level cameras for surveillance, industrial, and medical applications. In application, the sensor and other components are assembled on a printed circuit board (PCB). Due to the mismatch in the thermal expansion coefficient of the various materials, thermal deformation occurs on the PCB under the combined action of self-heating and thermal environment. As a result, the imaging quality of the sensor is reduced. Moreover, to acquire more image information under low-light conditions, the sensor is designed in a large format. Thus, the imaging quality is more susceptible to deformation. This issue has become a challenging problem in this field and needs to be solved urgently.

A robust optimization problem is considered for the camera module as Figure 6, in which the image-sensor-mounted PCB is fastened with the housing. The sensor is designed as a 4/3-inch optical format and features an array of five transistor pixels on a 6.5 μm pitch with an active imaging area of 2560(H) × 2160(V) pixels. It delivers extreme low light sensitivity with a read noise of less than 2.0 electrons root mean square (RMS) and a quantum efficiency above 55%. In order to analyze the thermal deformation of the sensor under the operating temperature (20 °C ∼ 45 °C), the finite element model (FEM) is created as shown in Figure 7, in which the power dissipation $\boldsymbol{P} = (P_1, P_2)$ of the codec chip and the converter is given as 1.2 W and 0.2 W, according to the test data. It can be observed that a certain deformation appears on the sensor die, and the peak–peak value (PPV) of the displacement response achieves about 3.0 μm. Consequently, the image quality of the sensor will decrease. To address the issue, the design objective is set to minimize the PPV, and the design variables of $\boldsymbol{X} = (X_1, X_2, X_3)$ are the normal positions of the PCB-fixed points. In engineering, manufacturing errors are unavoidable and power dissipation fluctuates with changing loads, and thereby $\boldsymbol{X}$ and $\boldsymbol{P}$ are treated as evidence variables. Their BPAs are summarized on the basis of limited samples, as listed in Table 2. This robust optimization is constructed as follows:

$$\min v, \max Bel(PPV(\boldsymbol{X}, \boldsymbol{P}) \le v) \tag{23}$$

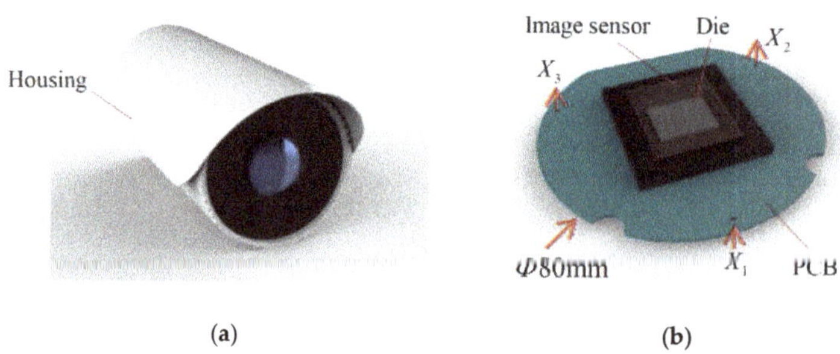


**Figure 6.** The camera module (**a**) with an ultra-low-noise image sensor (**b**). PCB: printed circuit board.

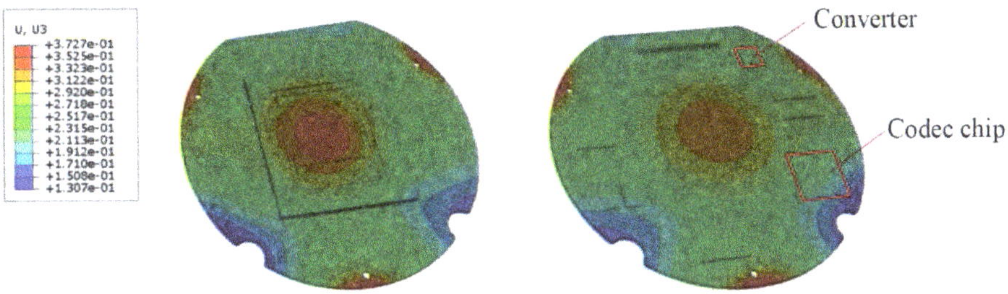


**Figure 7.** The finite element model (FEM) of the image-sensor-mounted PCB. PCB: printed circuit board.

**Table 2.** The marginal BPA of variables in the image sensor problem.

| $Xi, i = 1, 2, 3$ (mm) | | $P_1$ (W) | | $P_2$ (W) | |
|---|---|---|---|---|---|
| **Subinterval** | **BPA** | **Subinterval** | **BPA** | **Subinterval** | **BPA** |
| $[\overline{X}_i - 0.05, \overline{X}_i - 0.03]$ | 6.7% | [0.95, 1.1.05] | 37.5% | [0.55, 0.65] | 6.5% |
| $[\overline{X}_i - 0.03, \overline{X}_i - 0.01]$ | 24.2% | [1.05, 1.15] | 17.6% | [0.65, 0.75] | 23.8% |
| $[\overline{X}_i - 0.01, \overline{X}_i + 0.01]$ | 38.3% | [1.15, 1.25] | 5.4% | [0.75, 0.85] | 32.2% |
| $[\overline{X}_i + 0.01, \overline{X}_i + 0.03]$ | 24.2% | [1.25, 1.35] | 19.1% | [0.85, 0.95] | 12.5% |
| $[\overline{X}_i + 0.03, \overline{X}_i + 0.05]$ | 6.7% | [1.35, 1.45] | 20.4% | [0.95, 1.05] | 24.9% |

As mentioned above, the performance function of PPV is implicit and based on the time-consuming FEM, which consists of 88,289 8-node thermally coupled hexahedron elements. The computational time for solving the FEM is about 0.1 h, if using a computer with the i7-4710HQ CPU and 8 G of RAM. To realize the parameterization and reduce the computational cost of obtaining reference solutions, a second-order polynomial response surface is created for the performance function by sampling 200 times on the FEM.

$$\begin{aligned} PPV = {} & 2.396 - 9.924X_1 - 6.495X_2 + 14.178X_3 + 0.311P_1 - 0.226P_2 + 8.564X_1^2 + 16.960X_2^2 \\ & +14.104X_3^2 - 0.019P_1^2 + 0.794P_2^2 - 1.540X_1X_2 - 13.168X_1X_3 - 13.822X_2X_3 - 0.074P_1P_2 \end{aligned} \quad (24)$$

In order to analyze the efficiency of the proposed method for problems with different dimensional uncertainty, three cases are considered: only $\boldsymbol{P}$ is uncertain in Case 1; only $\boldsymbol{X}$ is uncertain in Case 2; and both of them are uncertain in Case 3. The initial design option is selected as $\overline{\boldsymbol{X}}^0 = (0.10\,\text{mm}, 0.10\,\text{mm}, 0.10\,\text{mm})$, and $v^0 = 2.92\,\mu$mis obtained by $PPV\left(\overline{\boldsymbol{X}}^0, \overline{\boldsymbol{P}}\right)$. After giving $Bel_j^T = (85\%, 95\%, 99.9\%)$, the original problem is converted into three sub-problems, as in Equation (9). They are solved by the proposed method and the double-loop method; all results are listed in Table 3. Firstly, the results of the proposed method and the reference solutions are almost identical for all cases, and thus the validity of the results is presented. Secondly, each of the cases converges into a stable optimal set after three or four iteration steps. For this problem, the convergence of the proposed method is little affected by the number of uncertain variables. Thirdly, the performance function evaluations $(N_F)$ increase with the increasing dimensional number of uncertainties, while overall the efficiency of the proposed method is relatively high. Case 3 is taken as an example. Even if the FEM is called directly by EBRO, $N_F = 198$ means a computational time of only about 20 h.

**Table 3.** The optimal set of the micro-force sensor problem.

| Results | | Case 1:2 Dimensions | Case 2:3 Dimensions | Case 3:5 Dimensions |
|---|---|---|---|---|
| Proposed method | $N_F$ | 128 | 142 | 198 |
| | Iterations | 3 | 4 | 4 |
| | $X_j^*$ (mm) | (0.200, 0.185, 0.000)<br>(0.200, 0.186, 0.000)<br>(0.200, 0.188, 0.000) | (0.200, 0.185, 0.000)<br>(0.200, 0.186, 0.000)<br>(0.200, 0.187, 0.000) | (0.200, 0.187, 0.000)<br>(0.200, 0.188, 0.000)<br>(0.200, 0.190, 0.000) |
| | $v_j^*, Bel_j^*$ | (1.35 μm, 85.1%)<br>(1.70 μm, 96.5%)<br>(2.00 μm, 100.0%) | (1.41 μm, 85.8%)<br>(1.55 μm, 96.4%)<br>(1.65 μm, 100.0%) | (1.74 μm, 86.4%)<br>(2.06 μm, 96.1%)<br>(2.62 μm, 100.0%) |
| Reference solution | $v_j^r, Bel_j^r$ | (1.35 μm, 85.1%)<br>(1.69 μm, 95.5%)<br>(2.00 μm, 100.0%) | (1.41 μm, 85.8%)<br>(1.55 μm, 96.4%)<br>(1.65 μm, 100.0%) | (1.71 μm, 85.8%)<br>(2.02 μm, 95.4%)<br>(2.56 μm, 99.9%) |

### 5.3. A Capacitive Accelerometer

The capacitive accelerometer [45] has become very attractive for high-precision applications due to its high sensitivity, low power consumption, wide dynamic range of operation, and simple structure. The capacitive accelerometer is not only the central element of inertial guidance systems, but also has applications in a wide variety of industrial and commercial problems, including crash detection for vehicles, vibration analysis for industrial machinery, and hovering control for unmanned aerial systems.

Most of the capacitive accelerometers consist of two main modules: the sensing structure and the signal processing circuit; the former plays a critical role in the overall product performance. The sensing structure in this example, as in Figure 8, mainly includes five parts: a fixed electrode, a movable electrode, a coil, a counter weight, block 1, and block 2. The material they are made of is listed in Table 4. The capacitance between the two electrodes varies with the vertical displacement of the movable plate under the excitation of acceleration, which can be clearly presented through the finite element simulation, as in Figure 9. The nodes in the effective area on the movable electrode have been offset relative to the original position under the excitation of acceleration. The increment of capacitance is expressed as [44]:

$$\begin{aligned} \Delta C &= \varepsilon \cdot A \\ A &= \sum_{i=1}^{n} \left( \frac{S_i}{h+\delta_i} - \frac{S}{h} \right) \end{aligned} \tag{25}$$

where ε is the dielectric constant; h represents the original distance between the electrodes; $S$ denotes the effective area on the movable electrode; $\delta_i$ is the displacement response of the $i$-th node, and $S_i$ is the area of corresponding element. Note that the performance function of $A$ is based on the FEM, which contains 114,517 8-node thermally coupled hexahedron elements in total, and it takes about 1/3 h to solve each time, when using a personal computer. The displacement of the movable electrode, in addition to the response to acceleration, may be caused by varying ambient temperature. This can be found from the simulation result in Figure 9, where the load is changed from acceleration to varying temperature. Reducing the effect of thermal deformation on accuracy has become a problem that must be faced in the design process. Therefore, the EBRO model of the capacitive accelerometer is formulated as:

$$\min v, \max Bel\left(f = \frac{A(\mathbf{X}, \boldsymbol{\alpha})}{\Delta T} \le v\right) \tag{26}$$

where $f$ represents the sensitivity of error to temperature at 35 °C; $v$ denotes the performance threshold; and $\mathbf{X}$, $\mathbf{P}$ denote the design vector and parameter vector. The components of $\mathbf{X}$, $\boldsymbol{\alpha}$ are the structural sizes, as shown in Figure 8, where $\alpha_1$, $\alpha_2$, $\alpha_3$ are the mounting angle of the counter weight, block 1 and block 2, respectively. The value ranges are given as $6.0\,\text{mm} \le \overline{X}_1, \overline{X}_2 \le 12.0\,\text{mm}$, and the nominal values are $\overline{\alpha}_i = 0\,\text{mm}$, $i = 1, 2, 3$. All of them are uncertain variables, respectively caused by machining errors and assembly errors. According to the existing samples, the marginal BPAs are listed in Figure 10.

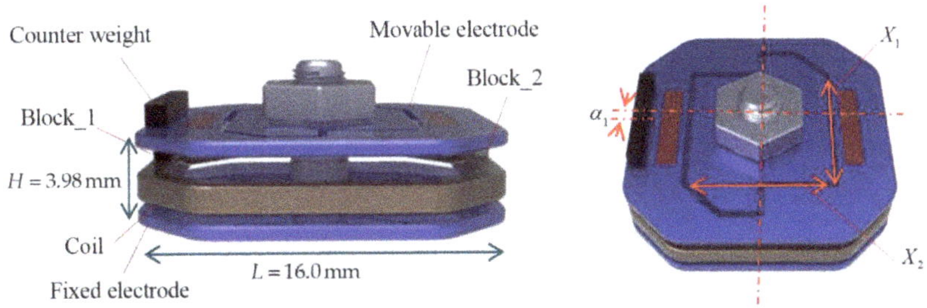


**Figure 8.** The sensing structure of a capacitive accelerometer.

**Table 4.** Material of accelerometer parts.

| Part | Fixed Electrode | Movable Electrode | Coil | Counter Weight | Block 1 | Block 2 |
|---|---|---|---|---|---|---|
| Material | Silicon | Silicon | Copper | Wolfram | Wolfram | Aluminium |

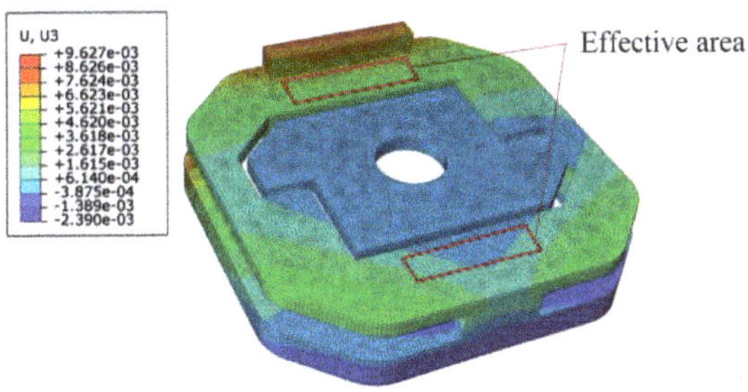


**Figure 9.** The FEM of the capacitive accelerometer.

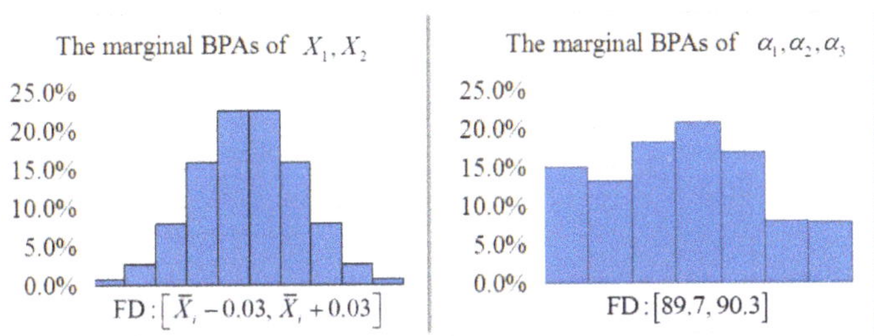


**(a)** **(b)**

**Figure 10.** The marginal BPAs of variables in the accelerometer problem. BPA: basic probability assignment; FD: frame of discernment.

After being given a series of $Bel_j^T = 80\%, 85\%, 90\%, 95\%, 99.9\%$, Equation (26) can be rewritten as Equation (27):

$$\left.\begin{array}{l} \min v \\ \text{st. } Bel\left(f = \frac{A(\mathbf{X},\boldsymbol{\alpha})}{\Delta T} \leq v\right) \geq Bel_j^T \end{array}\right\}, j = 1, 2, \ldots, 6 \quad (27)$$

For easy reproduction of results, the response surface of $A(\boldsymbol{X}, \boldsymbol{\alpha})$ is constructed as follows:

$$\begin{aligned} A = & 1.207X_1^2 - 0.430X_1X_2 - 18.06X_1 + 1.004X_2^2 - 13.974X_2 \\ & +100.9\alpha_1^2 - 9.0\alpha_1(\alpha_1 - 1) + 89.0\alpha_2^2 - 7.2\alpha_2 + 40.9\alpha_3^2 - 6.7\alpha_3 + 144.0 \end{aligned} \tag{28}$$

Next, the EBRO is performed by the proposed method and the double-loop method. The initial design point is selected as $\overline{\boldsymbol{X}}^{(0)} = (9.6\text{mm}, 9.6mm)$, and $f\left(\overline{\boldsymbol{X}}^{(0)}\right) = 0.565\mu\text{m}/°\text{C}$. All results are given in Table 5. The proposed method converges the optimal set after four iteration steps. Each element of the optimal set is very close to that of the reference solution. This indicates to some extent the convergence and accuracy of the proposed method. As for efficiency, the performance function evaluations of the proposed method are done 171 times. Compared to the double-loop method (12,842 times), the efficiency of this method has a definite advantage. Given that hundreds of simulations or dozens of hours of computation are acceptable for most engineering applications, it is feasible to directly call the time-consuming simulation model when performing the EBRO in practice.

**Table 5.** The optimal set of the accelerometer problem.

| Results | Proposed Method | Reference Solution |
|---|---|---|
| $\boldsymbol{X}_j^*$ (mm) | (9.0661, 8.9042) | (9.0661, 8.9042) |
| | (9.0659, 8.9040) | (9.0660, 8.9040) |
| | (9.0657, 8.9038) | (9.0657, 8.9039) |
| | (9.0653, 8.9035) | (9.0654, 8.9036) |
| | (9.0653, 8.9035) | (9.0648, 8.9031) |
| | (9.0653, 8.9035) | (9.0644, 8.9028) |
| $v_j^*, Bel_j^*$ | (0.244 μm, 81.8%) | (0.244 μm, 81.8%) |
| | (0.302 μm, 86.3%) | (0.279 μm, 85.2%) |
| | (0.350 μm, 90.8%) | (0.338 μm, 90.1%) |
| | (0.449 μm, 96.1%) | (0.424 μm, 95.5%) |
| | (0.607 μm, 99.4%) | (0.594 μm, 99.2%) |
| | (0.776 μm, 100.0%) | (0.733 μm, 99.9%) |

On the other hand, the proposed method provides six design options under different robustness requirements in this example. The higher the robustness, the greater the performance threshold. From a designer's point of view, choosing a lower yield (i.e., $Bel = 81.8\%$) means that a higher cost and less error (i.e., $v = 0.244\ \mu\text{m}/°\text{C}$) are introduced by temperature varying. Usually, the final design option is selected from the optimal set after balancing the cost and performance of the accelerometer. Objectively speaking, the proposed method does not provide a complete Pareto optimal set, but rather solutions under the given conditions. However, for the design of an actual product, the information of $Bel_j^T$ can usually be obtained on the basis of the engineering experience or the quality standards. Therefore, the proposed method is suitable for most product design problems.

## 6. Conclusions

Due to inevitable uncertainties from various sources, the concept of robust optimization has been deeply rooted in engineering designs. Compared to traditional probability models, evidence theory may be an alternative to model uncertainties in robustness optimization, especially in the cases of limited samples or conflicting information. In this paper, an effective EBRO method is developed, which can provide a computational tool for engineering problems with epistemic uncertainty. The contribution of this study is summarized as follows. Firstly, the improved EBDO model is formulated by introducing performance threshold as a newly-added design variable, and this model can handle the uncertainties involved in design variables and parameters. Secondly, the original ERBO is transformed into a series of sub-problems to avoid double-objective optimization, and thus the difficulty of solving is reduced greatly. Thirdly, an iterative strategy is proposed to drive the robustness analysis and the optimization solution alternately, resulting in nested optimization in the sub-problems achieving decoupling. The proposed method is applied to the three MEMS design problems, including a micro-force sensor, an image sensor, and a capacitive accelerometer. In the applications, the finite element simulation models and surrogate models are both given. Numerical results show that the proposed method has good engineering practicality due to comprehensive performance in terms of efficiency, accuracy, and convergence. Also, this work provides targeted engineering examples for peers to develop novel algorithms. In the future, the proposed method may be extended to more complex engineering problems with dynamic characteristics or coupled multiphysics.

**Author Contributions:** Conceptualization, Z.H. and S.D.; Data curation, J.X.; Formal analysis, J.X.; Funding acquisition, T.Y.; Investigation, Z.H.; Methodology, Z.H.; Project administration, T.Y.; Resources, S.D. and F.L.; Software, J.X.; Validation, S.D. and F.L.; Visualization, J.X.; Writing–original draft, Z.H.; Writing–review & editing, S.D.

**Funding:** This research was supported by the Major Program of National Natural Science Foundation of China (51490662); the Educational Commission of Hunan Province of China (18A403, 17A036, 17C0044); and the Natural Science Foundation of Hunan Province of China (2016JJ2012, 2017JJ2022, 2019JJ40296, 2019JJ40014).

**Conflicts of Interest:** The authors declare no conflict of interest.

## References

1. Taguchi, G.; Phadke, M.S. Quality Control, Robust Design, and the Taguchi Method. In *Quality Engineering through Design Optimization*; Springer: Berlin, Germany, 1989; pp. 77–96.
2. Beyer, H.G.; Sendhoff, B. Robust optimization—A comprehensive survey. *Comput. Methods Appl. Mech. Eng.* **2007**, *196*, 3190–3218.
3. Fowlkes, W.Y.; Creveling, C.M.; Derimiggio, J. *Engineering Methods for Robust Product Design: Using Taguchi Methods in Technology and Product Development*; Addison-Wesley: Reading, MA, USA; Bosten, MA, USA, 1995; pp. 121–123.
4. Gu, X.; Sun, G.; Li, G.; Mao, L.; Li, Q. A comparative study on multiobjective reliable and robust optimization for crashworthiness design of vehicle structure. *Struct. Multidiscip. Optim.* **2013**, *48*, 669–684. [CrossRef]
5. Yao, W.; Chen, X.; Luo, W.; van Tooren, M.; Guo, J. Review of uncertainty-based multidisciplinary design optimization methods for aerospace vehicles. *Prog. Aerosp. Sci.* **2011**, *47*, 450–479. [CrossRef]
6. Nguyen, A.T.; Reiter, S.; Rigo, P. A review on simulation-based optimization methods applied to building performance analysis. *Appl. Energy* **2014**, *113*, 1043–1058. [CrossRef]
7. Hoffman, F.O.; Hammonds, J.S. Propagation of uncertainty in risk assessments: The need to distinguish between uncertainty due to lack of knowledge and uncertainty due to variability. *Risk Anal.* **1994**, *14*, 707–712. [CrossRef] [PubMed]
8. Grubbs, F. An introduction to probability theory and its applications. *Technometrics* **1958**, *9*, 342. [CrossRef]
9. Gao, W.; Chen, J.J.; Sahraee, S. Reliability-based optimization of trusses with random parameters under dynamic loads. *Comput. Mech.* **2011**, *47*, 627–640.
10. Haldar, A.; Mahadevan, S.; Haldar, A.; Mahadevan, S. Probability, reliability and statistical methods in engineering design (haldar, mahadevan). *Bautechnik* **2013**, *77*, 379.
11. Chen, W.; Wiecek, M.M.; Zhang, J. Quality utility—A compromise programming approach to robust design. *J. Mech. Des.* **1999**, *121*, 179–187. [CrossRef]
12. Lee, K.H.; Park, G.J. Robust optimization considering tolerances of design variables. *Comput. Struct.* **2001**, *79*, 77–86. [CrossRef]
13. Zheng, J.; Luo, Z.; Jiang, C.; Gao, J. Robust topology optimization for concurrent design of dynamic structures under hybrid uncertainties. *Mech. Syst. Signal. Process.* **2018**, *120*, 540–559. [CrossRef]
14. Tzvieli, A. Possibility theory: An approach to computerized processing of uncertainty. *J. Assoc. Inf. Sci. Technol.* **1990**, *41*, 153–154. [CrossRef]
15. Georgescu, I. *Possibility Theory and the Risk*; Springer: Berlin, Germany, 2012.
16. Gupta, M.M. Fuzzy set theory and its applications. *Fuzzy Sets Syst.* **1992**, *47*, 396–397. [CrossRef]
17. Sun, B.; Ma, W.; Zhao, H. Decision-theoretic rough fuzzy set model and application. *Inf. Sci.* **2014**, *283*, 180–196. [CrossRef]

18. Elishakoff, I.; Colombi, P. Combination of probabilistic and convex models of uncertainty when scarce knowledge is present on acoustic excitation parameters. *Comput. Methods Appl. Mech. Eng.* **1993**, *104*, 187–209. [CrossRef]
19. Ni, B.Y.; Jiang, C.; Huang, Z.L. Discussions on non-probabilistic convex modelling for uncertain problems. *Appl. Math. Model.* **2018**, *59*, 54–85. [CrossRef]
20. Dempster, A.P. Upper and lower probabilities induced by a multivalued mapping. *Ann. Math. Stat.* **1967**, *38*, 325–339. [CrossRef]
21. Shafer, G. *A Mathematical Theory of Evidence*; Princeton University Press: Princeton, NY, USA, 1976.
22. Caselton, W.F.; Luo, W. Decision making with imprecise probabilities: Dempster-shafer theory and application. *Water Resour. Res.* **1992**, *28*, 3071–3083. [CrossRef]
23. Du, X. Uncertainty analysis with probability and evidence theories. In Proceedings of the 2006 ASME International Design Engineering Technical Conferences and Computers and Information in Engineering Conference, Philadelphia, PA, USA, 10–13 September 2006.
24. Vasile, M. Robust mission design through evidence theory and multiagent collaborative search. *Ann. N. Y. Acad. Sci.* **2005**, *1065*, 152–173. [CrossRef]
25. Croisard, N.; Vasile, M.; Kemble, S.; Radice, G. Preliminary space mission design under uncertainty. *Acta Astronaut.* **2010**, *66*, 654–664. [CrossRef]
26. Zuiani, F.; Vasile, M.; Gibbings, A. Evidence-based robust design of deflection actions for near Earth objects. *Celest. Mech. Dyn. Astron.* **2012**, *114*, 107–136. [CrossRef]
27. Hou, L.; Pirzada, A.; Cai, Y.; Ma, H. Robust design optimization using integrated evidence computation—With application to Orbital Debris Removal. In Proceedings of the IEEE Congress on Evolutionary Computation, Sendai, Japan, 25–28 May 2015.
28. Sentz, K.; Ferson, S. *Combination of Evidence in Dempster-Shafer Theory*; Sandia National Laboratories: Albuquerque, NM, USA, 2002; Volume 4015.
29. Jiang, C.; Zhang, W.; Han, X.; Ni, B.Y.; Song, L.J. A vine-copula-based reliability analysis method for structures with multidimensional correlation. *J. Mech. Des.* **2015**, *137*, 061405. [CrossRef]
30. Dong, W.; Shah, H.C. Vertex method for computing functions of fuzzy variables. *Fuzzy Sets Syst.* **1987**, *24*, 65–78. [CrossRef]
31. Jin, Y.; Branke, J. Evolutionary optimization in uncertain environments—A survey. *IEEE Trans. Evol. Comput.* **2005**, *9*, 303–317. [CrossRef]
32. Marler, R.T.; Arora, J.S. Survey of multi-objective optimization methods for engineering. *Struct. Multidiscip. Optim.* **2004**, *26*, 369–395. [CrossRef]
33. Park, G.J.; Lee, T.H.; Lee, K.H.; Hwang, K.H. Robust design: An overview. *AIAA J.* **2006**, *44*, 181–191. [CrossRef]
34. Huang, Z.L.; Jiang, C.; Zhang, Z.; Fang, T.; Han, X. A decoupling approach for evidence-theory-based reliability design optimization. *Struct. Multidiscip. Optim.* **2017**, *56*, 647–661. [CrossRef]
35. Zhang, Z.; Jiang, C.; Wang, G.G.; Han, X. First and second order approximate reliability analysis methods using evidence theory. *Reliab. Eng. Syst. Saf.* **2015**, *137*, 40–49. [CrossRef]
36. Breitung, K. Probability approximations by log likelihood maximization. *J. Eng. Mech.* **1991**, *117*, 457–477. [CrossRef]
37. Du, X.; Chen, W. Sequential optimization and reliability assessment method for efficient probabilistic design. *J. Mech. Des.* **2002**, *126*, 871–880.
38. Fletcher, R. *Practical Methods of Optimization*; John Wiley & Sons: Somerset, NJ, USA, 2013; pp. 127–156.
39. Coultate, J.K.; Fox, C.H.J.; Mcwilliam, S.; Malvern, A.R. Application of optimal and robust design methods to a MEMS accelerometer. *Sens. Actuators A Phys.* **2008**, *142*, 88–96. [CrossRef]
40. Akbarzadeh, A.; Kouravand, S. Robust design of a bimetallic micro thermal sensor using taguchi method. *J. Optim. Theory Appl.* **2013**, *157*, 188–198. [CrossRef]
41. Li, F.; Liu, J.; Wen, G.; Rong, J. Extending sora method for reliability-based design optimization using probability and convex set mixed models. *Struct. Multidiscip. Optim.* **2008**, *59*, 1–17. [CrossRef]
42. Fishman, G. *Monte Carlo: Concepts, Algorithms, and Application*; Springer Science & Business Media: Berlin, Germany, 2013; pp. 493–583.
43. Smits, J.G.; Dalke, S.I.; Cooney, T.K. The constituent equations of piezoelectric bimorphs. *Sens. Actuators A Phys.* **1991**, *28*, 41–61. [CrossRef]

44. Fossum, E.R.; Hondongwa, D.B. A review of the pinned photodiode for CCD and CMOS image sensors. *IEEE J. Electron. Devices Soc.* **2014**, *2*, 33–43. [CrossRef]
45. Benmessaoud, M.; Nasreddine, M.M. Optimization of MEMS capacitive accelerometer. *Microsyst. Technol.* **2013**, *19*, 713–720. [CrossRef]

*applied sciences*

MDPI

*Article*

# Improving Conductivity in Nano-Conduit Flows by Using Thermal Pulse-Induced Brownian Motion: A Spectral Impulse Intensity Approach

**Ugur Tuzun** [1,2]

[1] Chemical and Process Engineering Consultant; Weybridge KT13 9SF, UK; prof.ugur.tuzun@cantab.net
[2] Professor Alumnus, Churchill College, University of Cambridge; Cambridge CB3 ODS, UK

Received: 4 August 2019; Accepted: 8 September 2019; Published: 17 September 2019

**Featured Application: Nanoparticle suspensions are used in a variety of applications involving the transport of reagents and products to and from structured material surfaces. The more efficient alignment of particle layers adjacent to, and in contact with, a reactive surface, such as found in a fuel cell, bio-medical and electrochemical device, is often reliant upon the effective control of the dynamics of particle assembly in narrow conduits. One such control is possible by spectral thermal pulsing to generate a controlled Brownian motion of particles. This is demonstrated by numerical simulation here to be highly effective when particles are conveyed in viscous fluids close to a neutrally buoyant condition.**

**Abstract:** Inter-particle and particle-wall connectivity in suspension flow has profound effects on thermal and electrical conductivity. The spectral impulse generation and the imparting of kinetic energy on the particles is shown through a mathematical analysis to be effective as a means of achieving an approximate equivalent of a Langevin thermostat. However, with dilute suspensions, the quadratic form of the thermal pulse spectra is modified with a damping coefficient to achieve the desired Langevin value. With the dense suspension system, the relaxation time is calculated from the non-linear differential equation, and the fluid properties were supported by the viscosity coefficient. A "smoothed" pulse is used for each time-step of the flow simulation to take care of the near-neighbor interactions of the adjacent particles. An approximate optimal thermostat is achieved when the number of extra pulses introduced within each time step is found to be nearly equal to the co-ordination number of each particle within the assembly. Furthermore, the ratio of the particle kinetic energy and the thermal energy imparted is found to be never quite equal to unity, as they both depend upon the finite values of the pulse duration and the relaxation time.

**Keywords:** nanoparticle suspension; Brownian motion; spectral thermal pulsing; DEM simulations; Nano-device applications

## 1. Introduction

The shallow flows of nano-particles in planar conduits is a common feature of many industrial device applications, as well as being one of the primary generic routes used for transporting reagents and products to and from structured material surfaces as part of selective heterogeneous surface interactions, ranging from adsorptive separations to reaction catalysis; see for example (Pawar and Lee, 2015) [1], (Merino-Garcia et al. 2018) [2].

The more efficient alignment of particle layers adjacent and in contact with a reactive surface, such as found in fuel cell, bio-catalysis and electrochemical surface coating applications, is profoundly important to the efficiency of the processes required to generate, store and release renewable energy, as well as providing uniform surface scaffolds for varied environmental emission detection and separation devices. Previous work using pore-scale CFD simulations (see for example Crevacore et al. 2017) [3] has investigated the effects of varying the direction of the acceleration due to gravity on colloidal suspensions from orthogonal to parallel to the fluid flow direction, and the resulting impact on Brownian motion and particle settling.

Albert Einstein (1956) [4] formulated the mathematical framework for the kinetics of the Brownian motion of particles in suspension by considering the auto-correlation of fluctuating velocities and the spectral density of the intensity of fluctuations. It is possible to augment the physics of pressure-driven dense assembly flows of near-neutrally buoyant particles by combining the effects of (i) the lubrication forces due to the conveying fluid-immersed particle interactions and (ii) the drag forces due to fluid viscosity with those due to (iii) the thermal pulse-induced Brownian motion of particles.

In a series of DEM simulations of sub-micron Poiseuille flow in a narrow planar channel, the amelioration effect of the quadratic pulse intensity is demonstrated for different combinations of pulse intensity, relaxation time and time-step of simulations as a function of the solid fraction of the suspensions. Most significantly, it is possible to re-direct the clustering of particles in better alignment with the conduit wall surface when thermal pulses are applied. In the absence of pulses, the pressure-driven flows tend to favor a shear-induced migration of particles towards the center of the conduit, where the bulk shear gradient is reduced to zero due to symmetry; see (Davis et al. 2008) [5].

## 2. Modeling of Lubrication and Viscous Drag Forces

With dense particle suspensions, the forces between particle pairs are dominated by the lubrication limit, which holds when the gap width between particles, h, is much smaller than their mean diameter, $\dot{h}$. The interaction needs to be modified to take account of the roughness of the particle surfaces (no matter how small this may be) to avoid a singularity at h = 0. The interaction needs further augmentation to indicate what happens at h = 0, and for this eventuality a simple collision rheology is introduced as follows.

When two perfectly smooth particles in a fluid in slow flow come close together, the fluid mechanics may be approximated by the lubrication limit. In this limit, the interactive force for approach or departure normal to the smooth surfaces at the surface-to-surface distance h is proportional to $\dot{h}/h$. It is therefore never possible for perfectly smooth particles to have a solid contact (h = 0) at a finite velocity. In nature, physical surfaces are never perfectly smooth, and the interaction deviates from the ideal. A roughness dimension has to be introduced. In a paper by Jenkins and Koenders (2005) [6], it is argued that the rough surface effectively behaves as a permeable medium, so that the fluid in squeeze flow can escape by flowing parallel to the surface through the asperities on the surface. In this case a finite relative velocity is possible, and the interactive force behaves more like $\dot{h}/(h + h_0)$, where $h_0$ is a measure of the surface asperity size. Now it is possible to reach h = 0; the particles can touch.

For the purposes of the investigation, suspension flow is here approximated by the motion of an aggregate of neutrally buoyant particles of more or less equal diameter. Then the lubrication force $F_{12}$ is calculated by using the following equation;

$$F_{12} = -3\pi\eta D^2/8\,(h + h_0)\,.((v_1 - v_2).\hat{n}_{12})\,\hat{n}_{12} \quad (1)$$

where $\eta$ is the viscosity, $\hat{n}_{12}$ is the unit normal vector pointing from the surface of particle 1 to the surface of particle 2, $D$ is the diameter of the particles, $v_1$ and $v_2$ are the velocities of particles 1 and 2. The force mediated by the fluid in this way works purely normal to the surfaces, and a detailed discussion and derivation is in Vahid (2008) [7]. The tangential interaction is of an order less than the leading term in Equation (1) behaving as $\dot{h}\ln(h/D)$, see Jeffrey and Onishi (1984) [8].

When particles touch at a finite velocity, a collision mechanism is invoked. The normal velocity is updated using the coefficient of restitution, e using the unit normal vector

$$(v_2 - v_1)'.\,\hat{n}_{12} = -(1 - e)\,(v_2 - v_1).\,\hat{n}_{12} \quad (2)$$

$\overline{n}_{12}$ is introduced as the tangential vector such that $\overline{n}.\overline{n} = 0$. The tangential velocity is unaffected by the collision, so

$$(v_2 - v_1)^{/}.\,\overline{n}_{12} = (v_2 - v_1).\,\overline{n}_{12} \tag{3}$$

and momentum is conserved,

$$(m_1 v_1 + m_2 v_2) = m_1 v_2 + m_2 v_2 \tag{4}$$

These equations have been solved (with a symbolic manipulation program) for $v_1^{/}$ and $v_2^{/}$ and then implemented in the computer program.

In order to characterize the viscous drag forces in a particle suspension flow, two alternative approaches may be introduced: The "ensemble-averaged" option, based on a continuum consideration of the interstitial fluid, is to choose the width of the channel and the "suspension viscosity" that is the viscosity of the particle-fluid mixture. In general, the mixture viscosity is a solid fraction dependent factor greater than that of the interstitial fluid, see for example Thomas (1965) [9]. The second option is to choose the particle diameter, and the viscosity of the interstitial fluid thus then allows the calculation of a fluid-drag coefficient as a function of particle size and sphericity; see for example Seville et al. (1997) [10].

For suspension, Reynolds numbers close to Re = 1 (i.e., Stokes dynamics) of the suspension comprising near-neutrally buoyant particles, the effect of the fluid viscosity on the collisional dynamics of the particles can be estimated with the "ensemble-averaged" approach. One such estimate is provided by Davis et al. (2008) [5] where the following formula is derived for the shear viscosity of an isotropic medium;

$$\overline{\eta} = [3\eta\phi N_C/40]\left(D/\overline{h}\right) \tag{5}$$

Here, $\overline{\eta}$ is the suspension viscosity, $N_c$ is the average number of nearby neighbors ($N_c = 6$ is a good average for a dense suspension) and $D/\overline{h}$ is the ratio of particle diameter to mean surface-to-surface distance, and is a solid fraction-dependent quantity that was estimated by Torquato et al. (1990) [11];

$$\overline{h}/D = (1-\phi)^3/12\,\phi\,(2-\phi) \tag{6}$$

where the value of the solid fraction, $\phi$ typically ranges from 0.1 to about 0.5 through aggregation of dilute to dense suspensions; see for example Clayton et al. (2012) [12].

## 3. Characterization of Dem Simulation Results of Pressure-Driven Suspension Flows

The simulation runs for the external pressure-driven particulate suspension flow case are most easily characterized by means of the Péclet number (see also Morris and Boulay (1999) [13] and Frank, et al. (2003) [14], which is defined as

$$P_e = \frac{6\pi\eta\dot{\gamma}}{k_B T}\,\overline{r}^3 \tag{7}$$

where $\eta$ is the fluid viscosity, $\dot{\gamma}$ is the shear rate, $\overline{r}$ is the mean radius of the particles, $k_B$ is the Boltzmann constant and $T$ is the thermodynamic temperature.

In order to determine these parameters, the graphs of the profiles generated by the simulations are used. For $\dot{\gamma}$, the velocity profile is employed and an average is calculated from the maximum and minimum velocities and the width of the channel Hp;

$$\dot{\gamma} \cong 2(v_{max} - v_{min})/H_p \tag{8}$$

$k_B T$ is estimated from the equi-partition assumption and is found from the value of the kinetic energy profile at the center of the channel.

A number of 3D DEM simulation results is presented, both with and without Brownian motion for external pressure driven flow depicted by the schematics in section 4 below. It is shown that the simulation captures the migration effect very well, and shows that when the Brownian forces are increased (i.e., reduction of the Péclet number) the migration effect fades. In Figure 1a, the solid fraction profiles are calculated for different values of the Péclet number as a function of the dimensionless distance across the channel $y_p$ in terms of the number of particle diameters. When the Brownian motion is switched off; i.e., the Péclet number equals infinity, the case of the near parabolic solid fraction profile is recovered as evidence of the high degree of migration towards the center of the planar channel in Poiseuille flow. In contrast, Figure 1b shows the solid fraction profile for the Péclet number, $P_e = 3$, evaluated at different time steps for the external pressure driven flow, and it is observed that, as for the very low value of the Péclet number, the solid fraction profile is flatter and the migration effect disappears; see Figure 1b.

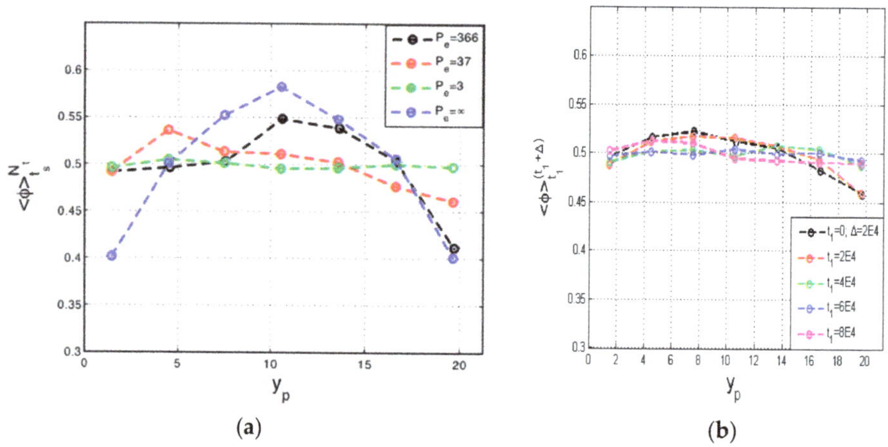


**Figure 1.** (**a**) Solid Fraction Profiles across the conduit at different Péclet numbers; (**b**) Solid Fraction Profiles with Péclet number = 3 at different simulation time-steps.

The effects of the presence and absence of the Brownian motion on the solid fraction profiles and the connectivity of the flowing particle assembly by the application of an external pressure for flow show quite clearly that the maintenance of a uniform (flat) solid fraction profile is possible only with a very low Péclet number, which according to Equation (7) also implies that the shear rate and the viscosity are quite low. Hence, in the dilute suspension case depicted here, the increased Brownian motion of the particles tends to encourage particle migration.

These results agree well with the near parabolic solid fraction and particle assembly velocity profiles reported earlier by Morris and Boulay (1999) [13] and Frank et al. (2003) [14]. Statistics of microscopic data, such as the particulate fluctuation energy, can also be presented from DEM simulation results; see Ibrahim (2012) [15].

## 4. Implementation of Brownian Motion by Thermal Pulses

For small particles in the sub-micron range, Brownian agitation introduces extra fluctuations, but this motion is thermally induced. While shear-induced fluctuations are position-dependent in flow in a conduit, the thermal fluctuations are position-independent (assuming an isothermal set-up), and therefore, when thermal fluctuations are greater or of the same order as the shear-induced ones, the migration pattern previously observed with shear-induced particle motion is diminished by this effect.

Brownian motion is implemented through a time-dependent, fluctuating force on the particles. Starting from dilute aggregates, a fluctuating force spectrum is derived, which embodies the properties of the fluid, and the momentum transfer from the fluid which can be carried over to the dense suspension case; see Ibrahim (2012) [15] for details. By adding fluctuations, the aggregate increases its energy, which has to be conducted away through the walls of the flow channel. The thermal conduction process is initiated by giving the walls a fixed temperature.

A one-dimensional problem is examined in the presence of thermal fluctuations, which were introduced to the simulation by the spectral intensity approach. The problem is studied for both cases, dilute and dense slurries. The results for the dilute case were compared with the Langevin method; see for example Davidchack et al. (2009) [16] and Koenders et al. (2012) [17]. Here, the quadratic velocity gives the same answer, while the quadratic displacement is not adequately reproduced. Therefore the quadratic equation is modified in the properties of the fluid in terms of the damping coefficient, and the Langevin value is achieved for small values of the pulse duration compared to the relaxation time; see Section 4.1 below.

The same exercise is done for the dense system where the relaxation is calculated from the non-linear differential equation, and the fluid properties were supported by the viscosity coefficient. It was also observed that the ratio of the kinetic energy and the thermal energy never equal to unity, as they depend on the ratio of the pulse duration and the relaxation time.

The recent work by Ishizuka et al. (2016) [18] also has demonstrated the use of pulsed gas flow in controlling the solids mass flow rate in the riser and downer sections of a triple-bed circulating fluidized reactor. The modulations of pulse width and pulse density (frequency) are used to optimize solids circulation rates by a combination of a low-pass filter and the electrical pulse voltage facilitated by a switching power supply.

### *4.1. Spectral Intensity of Thermal Energy Fluctuations*

In order to introduce thermal fluctuation to the suspension flow simulation, a spectral intensity approach of impulse distribution has been used on particles for a given fluid viscosity and temperature. The mean quadratic pulse intensity is dependent on the pulse duration and relaxation time, which is calculated by an auto-correlation function, whilst assuming that the pulse duration $\lambda$ is much smaller than the relaxation time.

The mean quadratic impulse value is given by;

$$< \mathrm{I}^2 > = \int \mathrm{I}^2 p(\mathrm{I}) d^2\mathrm{I} \;=\; -\frac{\beta}{\pi}\frac{\partial}{\partial\beta}\int e^{\beta I^2}\, d^2\mathrm{I} \;= 1/\beta \tag{9}$$

The value of $\beta$ is proportional to the mean kinetic energy imparted where $\beta = m/\left(2b^2\, k_B\, T\right)$ identified in two-dimensions for small values of the pulse duration, $\lambda$ by replacing the mean kinetic energy by $k_B T$, instead of $1/2\, k_B T$, (which is the one-dimensional value).

It also follows that b = $2\sqrt{2m\frac{\overline{\gamma}}{\lambda}}$, where $m/\overline{\gamma}$ is the mean relaxation time. Here, the pulse duration, $\lambda << m/\overline{\gamma}$.

To generate the impulse distribution, the angle is chosen to be entirely random. So, if the probability to encounter a value of I with components between $\mathrm{I}_1$ and $\mathrm{I}_1 + d\mathrm{I}_1$ and $\mathrm{I}_2$ and $\mathrm{I}_2 + d\mathrm{I}_2$ is

$$\mathrm{p}(\mathrm{I}_1, \mathrm{I}_2) \;=\; \frac{\beta}{\pi} e^{-\beta I^2} d\mathrm{I}_1 d\mathrm{I}_2, \tag{10a}$$

then

$$\mathrm{p}(\mathrm{I}, \theta) \;=\; \frac{\beta}{\pi} e^{-\beta I^2}\, \mathrm{I} d\mathrm{I}_1 d\theta \tag{10b}$$

Integrating over the angle then gives the distribution of the magnitude of the impulses yields;

$$\mathrm{p}(\mathrm{I}) \;=\; 2\,\frac{\beta}{\pi} e^{-\beta I^2} \mathrm{I} d\mathrm{I} \tag{11}$$

In practice, the impulses at very high values of I cannot be attained. In order to control this limitation, consider the values generated up to a maximum $I_{max}$. Integrating the frequency distribution in the interval $0 < I < I_{max}$ results in

$$\int_0^u P(I) = \int_0^u 2\beta I e^{-\beta I^2} = 1 - e^{bI^2_{max}} \tag{12}$$

The value of $I_{max}$ is chosen such that 99% of all values are covered. In practice values are stored in an array, according to frequencies that correspond to Equation (11), up to the value of $I_{max}$, which occurs only once. The angle is chosen at random between 0 and $2\pi$.

The simulation results can then be analyzed in terms of three key parameters: Pulse duration, relaxation time and simulation time step for particle motion.

### 4.2. Spectral Intensity Generation of Thermal Impulses in 1-D and 2-D Conduit Flows

Using the framework, which has been developed first for one-dimensional systems, the effects of thermal agitation are also introduced as a fluctuating force in a two-dimensional setting. Thus, the motion of a dense particle slurry that is driven through a conduit by a pressure, combined with Brownian motion, can be studied. The applications for such a system are obviously in nano-technology and in the separation and manipulation of systems in Chemical Engineering where small, sub-micron sized particles are present.

The simplest simulation is an aggregate in a box that is agitated by thermal motion; see Figure 2 below. There are fixed walls at the top and bottom, and periodic boundary conditions left and right. Initially, the wall particles in Figure 2 are given the same fluctuating velocity as the bulk particles; that is, the thermal motion is imparted to them. This allows them to acquire velocity whilst they are kept in the same position. Each particle in the bulk is given an impulse force from the distribution in Equation (11). The resulting particle distribution in a box of 200 particles is shown in Figure 3a.

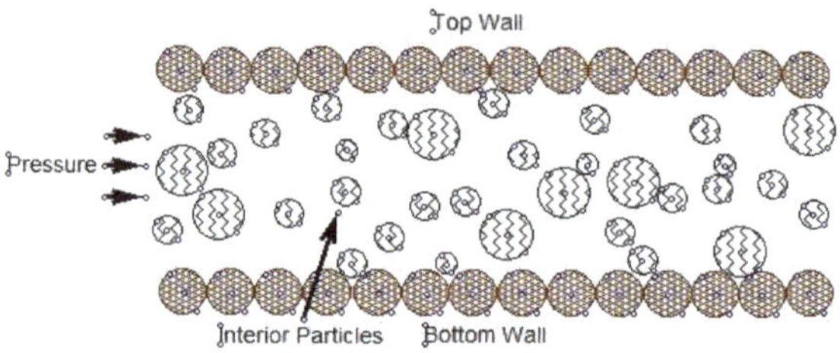


**Figure 2.** Simulation Channel for Particle-Fluid Suspension.

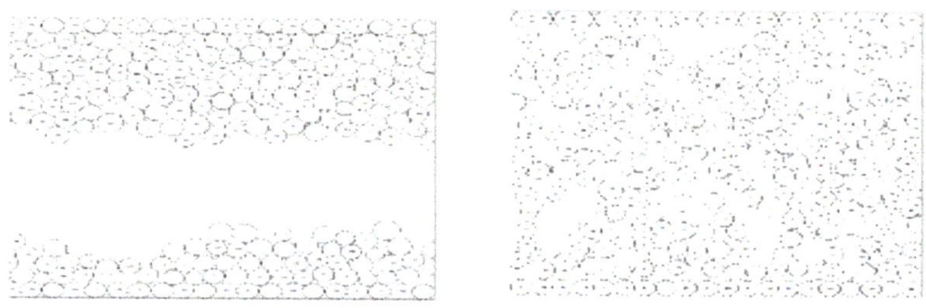

(a) (b)

**Figure 3.** (**a**) Particle Distribution after t = 50,000 time-steps with $k_BT$ = 3.5 $\times 10^{-3}$; (**b**) Particle Distribution with 'Cold' Walls with $k_BT$ -> 0 (t > 50,000).

If the walls are left completely "cold", the effect disappears, but a significant migration away from the walls appears as seen in Figure 3b. In order to experiment with the fluctuation velocity which the wall particles attain, the mass of the wall particles is varied. It transpires that for a mass ratio of about 20/1, a uniform (flat) voidage profile is achieved within the conduit. However, this is found to be not sustainable for larger time scales (t > 50,000). The particle migration effect can be suppressed by giving the particles an extra number of pulses in each time-step, so that the effect is a "smoothed" impulse. Further experimentation has established that if the number of extra pulses is equal to the number of nearby interacting particles, an approximate optimum is achieved; see for example; Koenders et al. (2012) [17].

These benchmark simulation experiments are believed to yield two very significant physical outcomes which can facilitate potentially a practical means of controlling the nano-particle agitation and particle separation in conduit flows. The effects suggest that (i) it is possible to impart controlled migration towards the center, and (ii) it is also possible to "clump" particles next to the wall and/or in a periodic cycle to alter the connectivity of the particle bed within the conduit. To achieve these effects in real-life, we would need to impart pulsed impulses to the walls whose materials can be chosen to yield a desired solid density ratio to the particles. A possible application is in the design and operation of electrochemical batteries used for power generation, where the interactions between the electrodes and the charge carrying particle bed can significantly effect the efficiency of electrical charge build-up and controlled release with minimal leakage; see Figure 4 below for a schematic illustration, and refer also to Kim et al. (2013) [19] and Bebelis et al. (2013) [20].

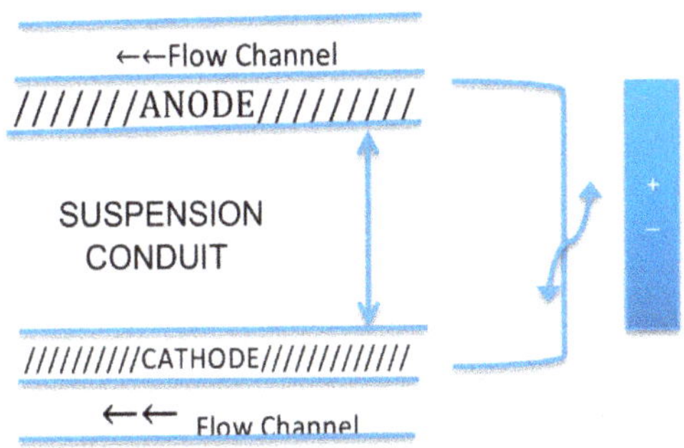


**Figure 4.** Schematic of a Suspension Flow Fuel Cell.

## 5. Practical Methods to Enhance and Manipulate Brownian Motion in Nano Particle Suspensions

The literature on nanoparticle suspension conductivity (for example Jang and Choi (2004) [21] and Keblinski et al. (2002) [22]) has considered four possible modes of energy transport in nanoparticle suspension flows: (1) The collisions between carrier fluid molecules, (2) thermal diffusion of nano-particles in suspension, (3) collisions between nanoparticles and (4) thermal interactions between dynamic nanoparticles and the carrier fluid molecules.

The earlier work has found that the temperature-dependent conductivity of nanoparticle suspensions at a specific concentration normalized to the conductivity of the carrier fluid can reveal increases by anything up to two-fold, depending on the fluid temperature. Jang and Choi (2004) [21] present results for $Al_2O_3$ particles in water and CuO particles in ethylene glycol, and show that the size of the nanoparticles has a very strong effect on the increase in normalized conductivity achieved. They report that when the fluid viscosity is relatively high like in ethylene glycol, reducing the nanoparticle size to < 10 nm, conductivity increases by two fold; however, when water is used as carrier fluid, the conductivity is reduced with decreasing nanoparticle size.

These results therefore suggest a sensitivity of the ratio of collisions between the nano-particle and carrier fluid molecules and the collisional frequency of the fluid molecules on the values of conductivity achieved. If we also take on board the assertion by Keblinski et al. (2002) [22] that the collisions between nanoparticles are an order of magnitude smaller than the collisions between the nanoparticle and fluid molecules, then the ratio of the pulse intensity to mean relaxation time as a key parameter (see Equation (9)), presented in the thermal pulse intensity spectra analysis presented above, gains further prominence supported by these earlier results reported on the conductivity of nanoparticle suspensions.

Keblinksi et al. (2005) [23] in a more recent study have also shown that the dynamic clustering and aggregation of nanoparticles is more pronounced as the volume concentration of nanoparticles exceeds 5%. This in turn results in a more rapid conductivity due to the increase in particle-particle contacts within the particle clusters. They report results of a several-fold increase in conductivity in consideration of the transformation from randomly closely-packed to loosely-packed spheres; see Keblinksi et al. (2005) [23]. This could only be due to the increased collisional activity of the nanoparticles in loosely-packed particle assemblies providing enhanced connectivity of the nanoparticle assembly.

The pulse intensity spectra analysis presented here shows how the balance between fluid-fluid molecule and fluid-nanoparticle interactions could be altered by affecting the Brownian motion of nanoparticles as function of fluid viscosity and nanoparticle size. Furthermore, it may be possible to make the water-based suspension to behave more like an ethylene glycol suspension, and vice versa, by the imparting of optimal frequency of pulsing for a given concentration and size of nanoparticles. Hence, according to the above analysis, the nanoparticle collisional frequency and its manipulation plays an important role in achieving the optimal value of the ratio of the rate of particle kinetic and thermal energy imparted. This in turn scales the macro-scale conductivity of the nanoparticle suspension relative to the molecular conductivity of the carrier fluid as defined by the well-known Maxwell Model; see Maxwell (1873) [24], which covers the original treatise on electricity and magnetism.

## 6. Comparison of Large-Scale Simulation Results

Figures 5 and 6 below compare the thermally pulsed assembly with that of the externally pressurized flow after steady-state is achieved within the flow channel in 60 k simulation time steps with 20,000 particles. The simulations were run for 120 k time steps to achieve accurate and reproducible results. This duration of flow simulation, on average, pumps the whole aggregate through the channel about 40 times, which corresponds to 12 s real-time.

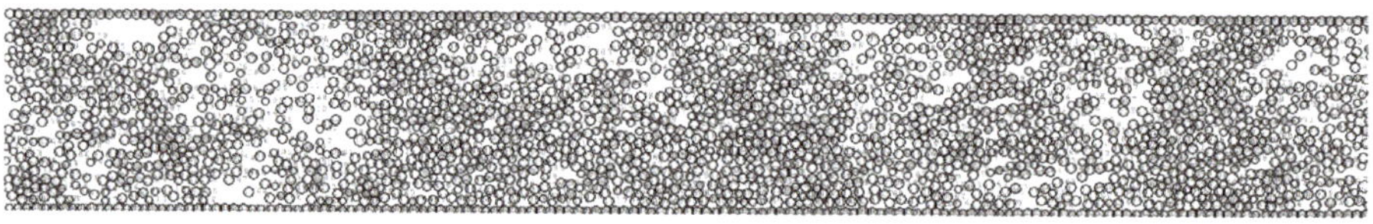

**Figure 5.** Thermally-pulsed assembly flow snapshot at time-step = 100 k.

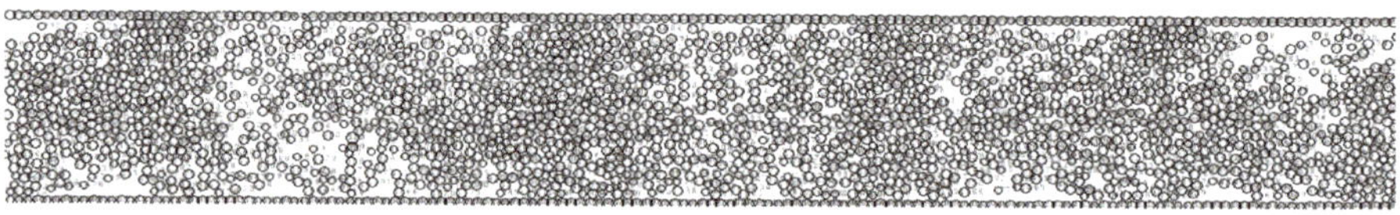

**Figure 6.** External Pressure driven flow of particulate assembly snapshot at time-step = 100 k.

In both the external-pressure driven flow and thermal-pulsed assembly driven flow scenarios, the suspension is advanced by the creation of periodic cavities adjacent to the planar channel walls coupled with the migration of particles towards the channel center.

There is the significance of the choice of the value of the coefficient of restitution, e = 0.2 (e.g., pharmaceutical compacts; see for example (Bharadwaj et al. 2010) [25] and the fluid viscosity of $\eta$ = 0.985 Pa-s (e.g., castor oil) with density similar to that of water in the simulation results presented here to highlight the interactions between highly inelastic "soft" collisions in the presence of moderately high fluid viscosity. These choices were made to allow the assembly simulations to reach "steady-state" within a reasonable computational time frame; see for example, Koenders, et al. (2001) [26].

These results are significant in illustrating the similarity of the nano-conduit flow physics obtained with the creation of Brownian motion in external pressure driven and thermal-pulsed conditions. The advantage of the latter is likely to be realized in applications allowing the energy transfer between the walls and the particle suspension, such as found in fuel cell and electrochemical battery applications. The current DEM simulations, if extended to longer real time flows and larger particle systems, can pave the way to our further understanding of the role of the particle-particle connectivity in particle beds under both Brownian motion-enhanced and -suppressed conditions. The ability to control the extent of Brownian motion in device flows such as seen in Figure 4 above may hold the key to mastering controlled energy storage and delivery with much greater efficiencies achieved to date; see for instance, Mateo et al. (2019) [27].

## 7. Conclusions

The results presented of the DEM simulations of neutrally-buoyant sub-micron particles in a flow regime where Re ≅ 0.8 (i.e., Stokes flow), concur well with the earlier observations made of channel flow experiments of suspension flows, whilst highlighting the effects of the controlled Brownian motion of particles via thermal pulsing. The amelioration of the shear-induced migration of particles in planar channel flows is demonstrated by the use of a DEM simulation framework that uses particle collisional dynamics in the presence of the fluid lubrication and viscous drag forces, which also allows for the implicit consideration of the surface roughness of particles through the $D/(h + h_0)$ ratio. The simulation results presented in Figures 5 and 6 demonstrate convincingly that the particle-particle and particle-wall thermal and electrical conduction processes are controlled effectively, and could be enhanced to suit practical purposes such as in electrode charging in fuel cells (Canizares et al. 2007) [28] and photovoltaic cells for solar energy storage (Mellor et al. 2016) [29] by manipulating the Brownian motion of the sub-micron particles. Conversely, the Brownian motion could be arrested to a measure to reduce the particle/wall friction experienced in micro-fluidic flows such as in the targeted introduction of bio-pharmaceuticals in the bloodstream, see Pele et al. (2015) [30].

The analytical framework of calculations presented here is capable of incorporation of short-range inter-particle attractive forces and gravitational acceleration to account for particle settling effects, and could be extended to flows in 3-D cylindrical tube configuration; see for example Crevacore et al. 2017) [3]. However, near-planar channel flows (2D) are also of significance in many microfluidic device applications. Most microfluidic devices rely on the operation of a peristaltic pump equivalent, and the thermal pulsing analysis presented here could be adapted to model the effects of pulse width and frequency modulation on collective particle motion in both surface (2D) and tubular (3D) flows; see Ishizuka et al. (2016) [18] for the latter.

**Funding:** Parts of this research received funding from Leverhulme Trust in early stages. Partial latter funding was provided by industrial consultancy activities in the U.S.A.

**Acknowledgments:** This research was supported by a European Federation of Chemical Engineering Energy Section Initiative. Private communications with Professor Manuel Andrés Rodrigo of the Dept. of Chemical Engineering, Faculty of Chemical Sciences and Technologies of the Universidad de Castilla La Mancha, Spain is gratefully acknowledged. Partial external funding was provided by industrial consultancy activities in the U.S.A. The funders had no role in the design of the study, in the collection, analyses and the interpretation of data; in the writing of the manuscript and in the decision to publish the results.

**Conflicts of Interest:** The author declares no conflict of interest.

## References

1. Pawar, R.C.; Lee, C.S. *Heterogeneous Nano-Composite Photocatalysis for Water Purification*; Micro & Nano Tech. Series; Elsevier Publ.: Amsterdam, The Netherlands, 2015; ISBN 978-0-323-39310-2.

2. Merino-Garcia, I.; Albo, J.; Kryzwda, P.; Guido, M.; Irabien, A. Bi-Metallic Cu-Based Hollow Fibre Electrodes for $CO_2$ Reduction. In Proceedings of the 6th International Conference on Green Process Engineering, Toulouse, France, 3–6 June 2018.
3. Crevacore, E.; Boccardo, G.; Grillo, A.; Marchisio, D.L.; Sethi, R. Pore-Scale Simulations of Particle Transport for Groundwater Remediation: The effect of Gravitational Settling. *Chem. Eng. Trans. (AIDIC)* **2017**, *60*, 193–198.
4. Einstein, A. *Investigations on the Theory of the Brownian Movement*; Dover: New York, NY, USA, 1956; ISBN 0-486-60304-0.
5. Davis, M.; Koenders, M.A.; Vahid, S. Granular Temperature Model for Oscillated Slurries: A Cell Model. *Proc. Inst. Mech. Eng. Part C J. Mech. Eng. Sci.* **2008**, *222*, 1995–2006. [CrossRef]
6. Jenkins, J.T.; Koenders, M.A. Hydrodynamics Interaction of Rough Spheres. *Granul. Matter* **2015**, *7*, 13–18.
7. Vahid, S. Migration and Structure Formation in Sheared Slurries. Ph.D. Thesis, University of Kingston, Kingston, UK, 2008.
8. Jeffrey, D.J.; Onishi, Y. Calculation of the Resistance and Mobility Functions for Two Unequal Rigid Spheres in Low Reynolds Number Flow. *J. Fluid Mech.* **1984**, *139*, 261–290. [CrossRef]
9. Thomas, D.G. Transport Characteristics of Suspension: A note on the viscosity of Newtonian suspension of Uniform Spherical Particles. *J. Colloid Sci.* **1965**, *20*, 267–277. [CrossRef]
10. Seville, J.P.K.; Tuzun, U.; Clift, R. *Processing of Particulate Solids*; Blackie Academic & Professional-Powder Technology Series; Chapman & Hall: London, UK; Kluwer Publ.: Amsterdam, The Netherlands, 1997; ISBN 978-0-7514-0376-8.
11. Torquato, S.; Lu, B.; Rubenstein, K. Nearest Neighbor Distribution Functions in Many-Body System. *Phys. Rev. A* **1990**, *41*, 2059–2071. [CrossRef]
12. Clayton, T.C.; Schwarzkopf, J.D.; Sommerfeld, M.; Tsuji, Y. *Multiphase Flows with Drops and Particles*; CRC Press, Taylor& Francis Group, LLC: Boca Raton, FL, USA, 2012; ISBN 978-1-4398-4050-4.
13. Morris, J.F.; Boulay, F. Curvilinear Flows of Non-Colloidal Suspensions: The Role of Normal Stresses. *J. Rheol.* **1999**, *43*, 1213–1237. [CrossRef]
14. Frank, M.; Anderson, D.; Weeks, E.R.; Morris, J.F. Particle Migration in Pressure-Driven Flow of Brownian Suspension. *J. Fluid Mech.* **2003**, *493*, 363–378. [CrossRef]
15. Ibrahim, M. Migration in Slow Slurry Flow including the Effect due to Brownian Motion. Ph.D. Thesis, University of Surrey, Guildford, UK, 2012.
16. Davidchack, R.L.; Handel, R.; Tretyakov, M.V. Langevin Thermostat for Rigid Body Dynamics. *J. Chem. Phys.* **2009**, *130*, 234101. [CrossRef]
17. Koenders, M.A.; Ibrahim, M.; Vahid, S. Chapter: DEM Simulation of Migration Phenomena in a Slow Dense Slurry. In *Discrete Element Modeling of Particulate Media*; RSC Publ.: London, UK, 2012; pp. 39–45, ISBN 978-1-84973-360-1.
18. Ishizuka, M.; Mizuno, H.; Kansha, Y.; Tsutsumi, A. Control of Solid Mass Flow Rate in Circulating Fluidised Bed by Pulsed Gas Flow. *Chem. Eng. Trans.* **2016**, *52*, 127–132.
19. Kim, H.; Boysen, D.A.; Newhouse, J.M.; Spatocco, B.L.; Chung, B.; Burke, P.J.; Bradwell, D.J.; Jiang, K.; Tomaszowska, A.A.; Wang, K.; et al. Liquid-Metal Batteries: Past, Present and Future. *Chem. Rev.* **2013**, *113*, 2075–2099. [CrossRef] [PubMed]
20. Bebelis, S.; Bouzek, K.; Cornell, A.; Ferreira, M.G.S.; Kelsall, G.H.; Lapicque, F.; Ponce de Leon CRodrigo, M.A.; Walsh, F.C. Ion Exchange Membranes: Fundamentals and Applications. *Chem. Eng. Res. Des.* **2013**, *91*, 1998–2020. [CrossRef]
21. Jang, S.P.; Choi, S.U.S. Role of Brownian Motion in the Enhanced Thermal Conductivity of Nanofluids. *Appl. Phys. Lett.* **2004**, *84*, 4316–4318. [CrossRef]
22. Keblinski, P.; Phillpot, S.R.; Choi, S.U.S.; Eastman, J.A. Mechanisms of Heat Flow in Suspensions of Nano-Sized Particles (Nanofluids). *Int. J. Heat Mass Transf.* **2002**, *45*, 855–863. [CrossRef]
23. Keblinski, P.; Eastman, J.A.; Cahill, D.G. Nanofluids for Thermal Transport. *Mater. Today* **2005**, *8*, 36–44. [CrossRef]
24. Maxwell, J.C. *A Treatise on Electricity and Magnetism*; Clarendon Press: Oxford, UK, 1873.
25. Bharadwaj, R.; Smith, C.; Hancock, B.C. The Coefficient of Restitution of Some Pharmaceutical Tablets/Compacts. *Int. J. Pharm.* **2010**, *402*, 50–56. [CrossRef]

26. Koenders, M.A.; Gaspar, N.; Tuzun, U. The Physical Effects of Structures Formation in Granular Materials. *Phys. Chem. Earth A Solid Earth Geod.* **2001**, *26*, 75–82. [CrossRef]
27. Mateo, S.; Mascia, M.; Fernandez-Morales, F.J.; Rodrigo, M.A.; Lorenzo, M.D. Assessing the Impact of Design Factors on the Performance of Two Miniature Microbial Fuel Cells. *Electrochim. Acta* **2019**, *297*, 297–306. [CrossRef]
28. Canizares, P.; Jiménez, C.; Martinez, F.; Sáez, C.; Rodrigo, M.A. Study of the Electrocoagulation Process using Aluminium and Iron Electrodes. *Ind. Eng. Chem. Res.* **2007**, *46*, 6189–6195. [CrossRef]
29. Mellor, A.; Hylton, N.P.; Hauser, H.; Thomas, T.; Lee, K.H.; Al-Saleh, Y.; Giannini, V.; Braun, A.; Loo, J.; Vercurysse, D.; et al. Nanoparticle Scattering for Multijunction Solar Cells: The Tradeoff between Absorption Enhancement and Transmission Loss. *IEEE J. Photovolt.* **2016**, *6*, 1678–1687. [CrossRef]
30. Pele, L.C.; Thoree, V.; Bruggraber, S.F.A.; Koller, D.; Thompson, R.P.H.; Lomer, M.C.; Powell, J.C. Pharmaceutical/food grade $TiO_2$ particles are absorbed into the Bloodstream of Human Volunteers. *Part. Fibre Toxicol.* **2015**, *12*, 26. [CrossRef] [PubMed]

applied sciences

MDPI

Article

# Effect of Soil Reinforcement on Tunnel Deformation as a Result of Stress Relief

**Huasheng Sun * and Wenbin Sun**

Huaiyin Institute of Technology, 89 North Beijing Road, Huai'an 223001, China; sunwb1969@163.com
* Correspondence: sunhuadasheng@126.com; Tel.: +86-18262811276

Received: 23 January 2019; Accepted: 29 March 2019; Published: 4 April 2019

**Abstract:** Adjacent geotechnical engineering activities, such as deep excavation, may adversely affect or even damage adjacent tunnels. Ground reinforcement before excavation may be an effective approach to reduce tunnel heave as a result of stress relief. However, there are few quantitative studies on the effect of soil reinforcement on tunnel deformation. Moreover, the reinforcement mechanism of the reinforced soil and the reinforcement depth are not fully understood. In order to investigate the effect of reinforcing the ground on the tunnel response, a finite element analysis was conducted based on a previously reported centrifugal model test with no ground reinforcement. The effect of the Young's modulus and depth of the reinforced soil on tunnel deformation was analyzed. Soil stresses around the tunnel were also considered to explain the tunnel response. The results revealed that the Young's modulus of the reinforced soil and the reinforcement depth had a significant impact on tunnel deformation as a result of basement excavation. The tunnel heave in the longitudinal direction decreased by 18% and 27% for modulus of the reinforced soil, five times and ten times higher than that of the non-reinforced soil, respectively. The reinforcement depth was effective with regard to controlling the tunnel heave caused by stress relief. This is because the reinforced soil blocked the stress transfer and thus reduced the tunnel heave caused by excavation unloading. It is expected that this study will be useful with regard to taking effective measures and ensuring the safety and serviceability of existing metro tunnels during adjacent excavation.

**Keywords:** basement bottom reinforcement; reinforcement depth; Young's modulus of reinforced soil; tunnel heave; numerical analysis

## 1. Introduction

In congested urban areas, the metro plays an important role in the transportation system. The safety and serviceability of existing metro tunnels are always under serious consideration. Owing to the rapid development of underground space in urban areas, the commercial demand for the construction of underground structures in close proximity to metro tunnel lines has been increasing. However, adjacent geotechnical engineering activities, such as deep excavation, may have adverse effects on or even cause damage to nearby existing tunnels. If the induced tunnel deformation and internal forces exceed the capacity of the tunnel structures, segment cracking, leakage, and even longitudinal distortion of the railway track may occur and seriously threaten the smooth travel and safety of the trains in operation. Many studies have investigated the effects of adjacent excavation on existing shield tunnels using various methods, including in situ monitoring [1–3], centrifuge model tests [4–6], numerical analysis [7–15], and semi-analytical methods [16–19]. For example, the main objects of investigation have been the excavation dimension [13,15], relative distance between the tunnel and excavation [4,8,9,11,13], construction and reinforcement methods [10,12], tunnel dimension and physical parameters [9,13], soil density and wall stiffness [6], different constitutive models [14], and influence zone [20].

Chang et al. [2] published an extensive report regarding the damage case history of a shield tunnel as the result of an adjacent deep excavation. In this case history, cracks in segmental linings and the distortion of connected blots were observed. Therefore, it is a major challenge for city designers and geotechnical engineers to evaluate and control shield tunnel responses associated with adjacent excavations. Although various construction and reinforcement methods of controlling tunnel deformation have been proposed by Hu et al. [10] and Liu et al. [12], the existing quantitative research on the evaluation of the bottom reinforcement is insufficient. Moreover, the effect of the soil reinforcement's mechanism on the tunnel deformation has not been clarified.

On the basis of a centrifugal model test reported by Ng et al. [4], the finite element simulation of the studied problem without ground reinforcement has already been done. Ground reinforcement before excavation may be a fresh approach to reduce tunnel heave as a result of stress relief. However, its influence law and deformation mechanism are not clear yet. On the basis of the centrifuge model test without ground reinforcement and the calibrated constitutive models reported by Ng et al. [14], a finite element analysis was conducted to investigate the effect of ground reinforcement on tunnel deformation, both in the longitudinal and in the transverse directions. The effect of the reinforced soil's Young's modulus and the effect of the reinforcement depth under the basement bottom were investigated. The effect of the basement excavation on the deformation of an existing tunnel was analyzed. It is expected that this study may deepen the understanding of tunnel deformation control methods and provide an effective method to control tunnel deformation due to excavation. This study will help in taking effective measures to ensure the safety and serviceability of existing metro tunnels during adjacent excavation for engineering applications.

## 2. Material and Methods

### 2.1. Description of the Simulated Centrifuge Test

The basis of the numerical simulation was derived from a centrifugal model test conducted by Ng et al. [4]. The tunnel was located just below the center of a pit, as shown in Figures 1 and 2. The centrifuge model test was conducted at the Hong Kong University Science and Technology centrifuge [21–23]. The foundation pit excavation was simulated using the discharge-of-heavy-liquid method, and the excavation depth was controlled using a piezometer. The excavation of the basement was carried out in three steps, and each excavation step was 3 m. Details regarding the process of the centrifuge model tests can be found in the report by Ng et al. [4].

### 2.2. Finite Element Analysis

#### 2.2.1. Finite Element Mesh and Boundary Conditions

The finite element program ABAQUS [24] was used to simulate the effect of the basement excavation on the existing tunnel. Figure 3 shows the three-dimensional finite element mesh adopted in the analysis. The mesh dimensions were 1200 mm in length, 990 mm in width, and 750 mm in depth. An eight-node brick element was used to simulate the sand and the diaphragm wall, and a four-node shell element was used to simulate the tunnel. Pin supports were applied to all vertical sides and the base of the mesh to restrain movement in any direction ($x$-, $y$- or $z$-direction).

In all numerical analyses, interface elements were used at the soil–tunnel and soil–basement wall interfaces, unless otherwise specified. Each interface element was a zero-thickness slip element governed by the Coulomb friction law. The friction coefficient ($\mu$) and limiting relative displacement ($\gamma_{\text{lim}}$), where the slippage occurred, were controlled by two input parameters for each slip element. The interface friction coefficient $\mu$ was derived from $\mu = \tan\delta$, where $\delta$ is the interface friction angle, which was taken as 20° (i.e., 2/3 of the soil's critical friction angle). The limiting displacement of 5 mm was assumed to achieve the full mobilization of the interface friction.

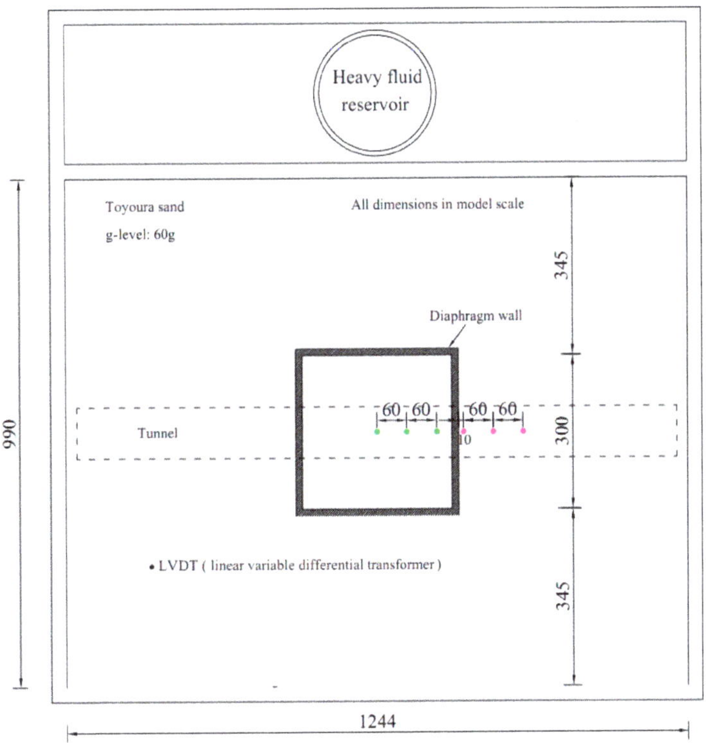


**Figure 1.** Plan view of the centrifuge model (redraw from Ng et al., 2013) [4].

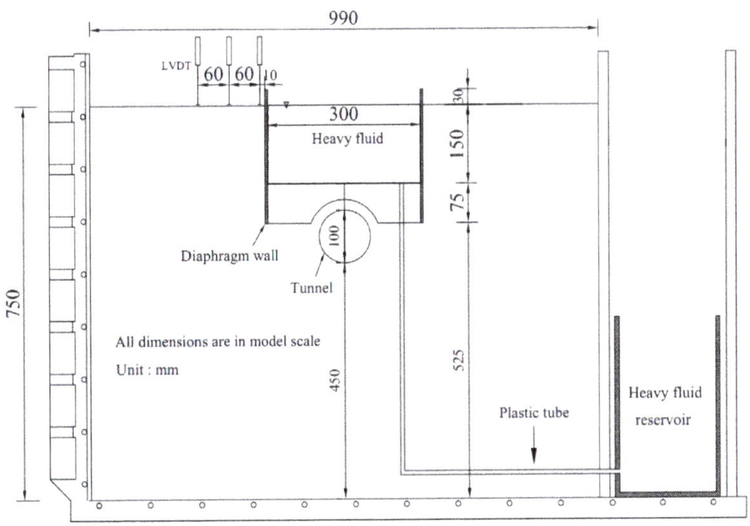


**Figure 2.** Elevation view of the centrifuge model in model scale (mm) (redraw from Ng et al., 2013) [4].

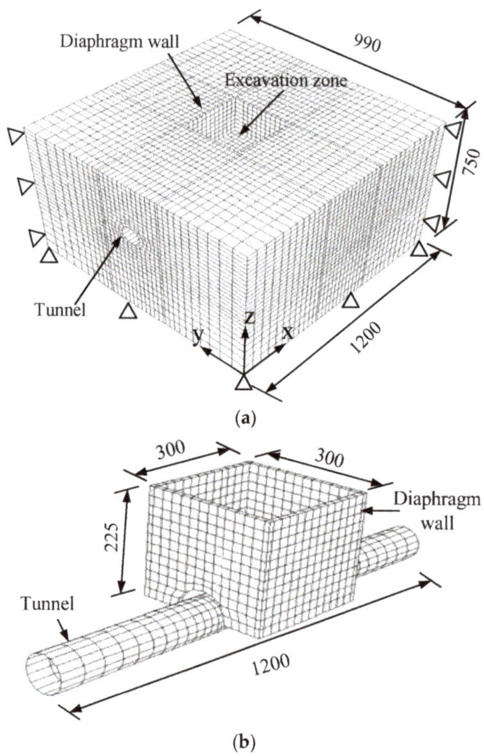


**Figure 3.** (**a**) Three-dimensional finite element model in model scale (mm) and (**b**) intersection of tunnel and diaphragm wall.

### 2.2.2. Constitutive Model and Model Parameters

In the experiment, the tunnel and diaphragm wall were made of aluminum alloy and considered elastic materials in the numerical analyses. The Young's modulus and Poisson's ratio of the aluminum alloy were 70 GPa and 0.2, respectively. The sand was simulated using the Mohr–Coulomb model as a first-order model based on the comparison reported by Ng et al. [14]. To describe the mechanical behavior of the soil in this model, five parameters are needed: Young's modulus $E$, Poisson's ratio $\nu$, friction angle $\varphi$, cohesion $c$, and dilatancy angle $\psi$. The chosen parameters are listed in Table 1. The parameter selection process and parameter values have been reported by Ng et al. [14].

**Table 1.** Soil parameters used in the Mohr–Coulomb model.

| Soil Parameters | Values |
|---|---|
| Young's modulus, $E_s$ | 117 MPa |
| Poisson's ratio, $\nu$ | 0.3 |
| Peak friction angle, $\varphi_p$ | 35° |
| Critical friction angle, $\varphi_c$ | 30° |
| Cohesion, $c$ | 2 kPa |
| Dilatancy angle, $\psi$ | 6° |

### 2.2.3. Numerical Modeling Procedure

The numerical modeling procedure is summarized as follows and shown in Figure 4:

1. Establish the initial stress conditions using $K_0 = 0.5$. Apply the same amount of vertical and horizontal pressure as that in the centrifuge test to the formation level and diaphragm wall, respectively.
2. Increment the gravitational acceleration of the entire model from 1 g to 60 g. Simultaneously, apply pressure to the formation level and wall.
3. Decrease the amount of vertical and horizontal pressure gradually in each excavation stage to simulate excavation up to a depth of 9 m.

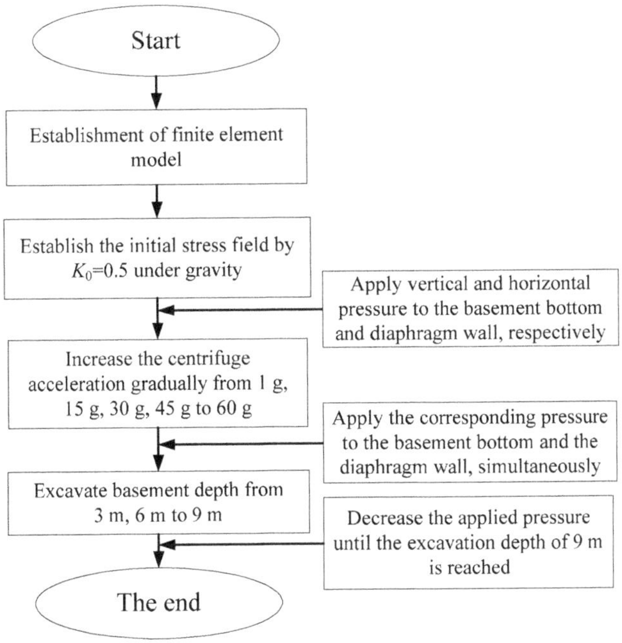


**Figure 4.** A sketch of the modeling procedure.

### 2.2.4. Numerical Modeling Scheme

It should be noted that there is no ground reinforcement in the centrifuge model test. However, the numerical modeling schemes include the cases with and without ground reinforcement. The centrifuge tests provide a benchmark for verifying the correctness and reasonability of the numerical results with no reinforcement. Furthermore, testing of the effect of the reinforcement on tunnel response due to stress relief can be conducted. The ground is assumed to be reinforced by grouting. All parameters of the ground mechanical behavior change after reinforcement may affect tunnel response. A parametric study was conducted to investigate the effect of soil friction angle, cohesion, etc., on the tunnel response. It was found that the strength indexes had only a little influence on tunnel deformation. However, the Young's modulus may play an important role in predicting the tunnel response. Thus, this study focused on the study of the soil's Young's modulus. Figure 5 shows the schematic diagram of the basement bottom reinforcement. The soil mass between the basement bottom and the tunnel was divided into parts 1, 2, and 3, with each part being 10 mm in the model (0.6 m in the prototype). Note that reinforcement S1 was equal to area 1, S2 was equal to area 1 plus area 2, and S3 was the sum of areas 1, 2, and 3. In the first scheme, the effect of the reinforced area's Young's modulus on the behavior of the tunnel was investigated, assuming that the reinforcement area underneath the basement was S1. The reinforced soil's Young's modulus varied between 1, 5, and 10 times the value of the soil's initial Young's modulus. As for the 5-times and 10-times values, they may not be close to reality but are certainly helpful for practice engineers to judge the effect of soil Young's modulus on the tunnel response. In the second scheme, the Young's modulus of the reinforced soil was constant ($E_r = 5E_s$), and the effect of the reinforcement depth on the tunnel's behavior was investigated. The reinforcement depth varied from area S1 = 0.6 m and S2 = 1.2 m to S3 = 1.8 m in the prototype.

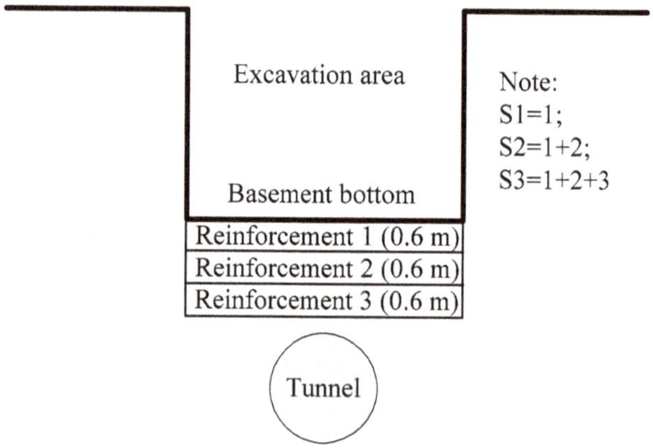


**Figure 5.** Reinforcement scheme for basement bottom.

## 3. Results and Analysis

### 3.1. *Effect of Reinforced Soil's Young's Modulus on Tunnel Heave*

Figure 6 shows the comparison of the measured and computed normalized tunnel heave and maximum tunnel heave variation with different soil reinforcement moduli, where $H_e$ is the final excavation depth. As can be seen in Figure 6a, the computed values were generally in agreement with the measured data reported by Ng et al. (2013) [4]. The difference between the measurements and the simulation results may be due to the chosen constitutive model. The adopted elastic–plasticity model with Mohr–Coulomb criterion could not capture the strain- and path-dependent soil stiffness. The maximum tunnel heave occurred at the tunnel crown underneath the basement center and decreased gradually as the distance away from the center of the basement increased.

According to the Land Transport Authority of Singapore [25], the maximum tunnel movement should not exceed 15 mm (i.e., 0.17% $H_e$). When none of the soil elements were reinforced, the computed maximum tunnel heave was 0.116% $H_e$, which overestimated the measured tunnel heave by 50%. In this study, both the computed and measured maximum tunnel heave were within the recommended allowable limit. Because the soil area S1 was reinforced with a Young's modulus that varied between $E_s$ and $5E_s$, the computed tunnel heave decreased greatly from 0.116% $H_e$ to 0.095% $H_e$, which overestimated the tunnel heave by 32%. However, the measured tunnel heave was overestimated by 23% as the reinforced soil's Young's modulus increased to $10E_s$. Thus, it is obvious that soil reinforcement underneath the basement reduced the tunnel heave. Further increase of the reinforcement's Young's modulus reduced the tunnel heave, but with a decreasing tendency. For comparison, the overestimated maximum tunnel heave is listed in Table 2. The basic value is the result obtained from the centrifuge model test.

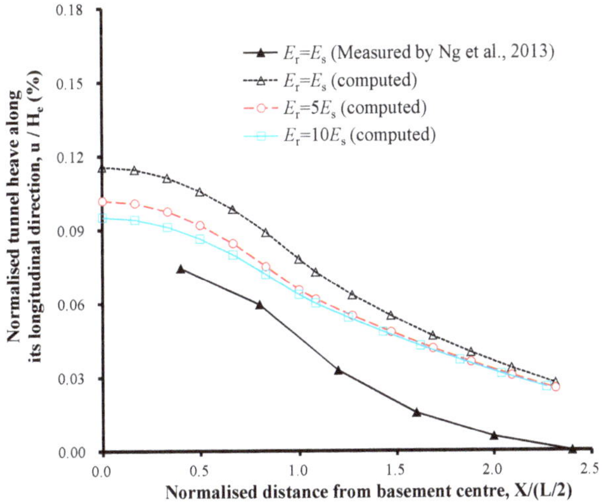


(a)

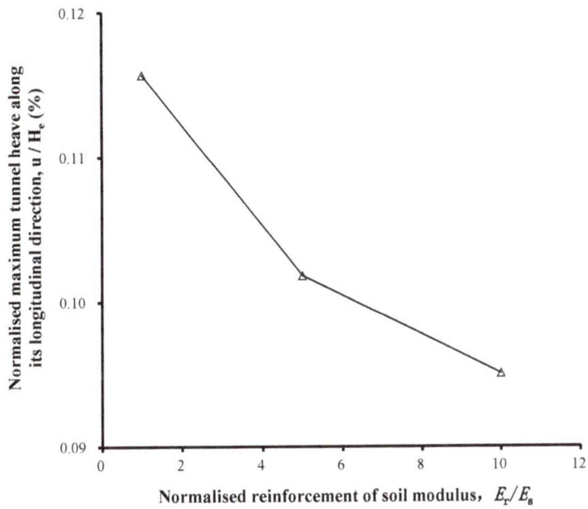


(b)

**Figure 6.** Measured and computed (**a**) tunnel heave and (**b**) maximum tunnel heave with different reinforced soil Young's moduli.

**Table 2.** Comparison of computed and measured values.

| Cases | Young's Modulus of Reinforced Soil, $E_r$ | Max. Tunnel Heave | Max. Change in Tunnel Diameter |
|---|---|---|---|
| No reinforcement | $E_r = E_s$ | +50% | −35% |
| Reinforcement S1 | $E_r = 5E_s$ | +32% | −39% |
| Reinforcement S1 | $E_r = 10E_s$ | +23% | −39% |
| Reinforcement S2 | $E_r = 5E_s$ | +31% | −39% |
| Reinforcement S3 | $E_r = 5E_s$ | +28% | −40% |

Note: The −ve sign indicates an underestimation and the +ve sign indicates an overestimation.

### 3.2. Effect of Reinforced Soil's Young's Modulus on Tunnel Diameter Change

Figure 7 shows the measured and computed change in the tunnel diameter with different soil reinforcement moduli. The positive and negative values denote the elongation and compression of the tunnel, respectively. It is noted that the described tunnel deformation was located in the tunnel section in a circumferential direction. All three cases predicted that the tunnel lining was vertically elongated and horizontally compressed and that the magnitude of the elongation ($\Delta D_V$) and compression ($\Delta D_H$) increased with the excavation depth. All of the computed values underestimated the tunnel diameter change in comparison to the experimental measurements. This can be attributed to the fact that, in the transverse direction, the computed soil stiffness around the tunnel was larger than that in the centrifuge model test. According to the British Tunnelling Society [26], the maximum distortion of a tunnel $((\Delta D_V + \Delta D_H)/D)$ should not exceed 2%. The maximum distortion of the existing tunnel (i.e., 0.09% $D$), which was induced by the basement excavation in this study, was within the abovementioned allowable limit. The increase of the reinforced soil's Young's modulus varied from $E_s$ to $5E_s$, which led to the maximum elongation of the tunnel lining of 9%. Significant changes were not observed when the reinforced soil's Young's modulus increased from $5E_s$ to $10E_s$. Thus, it was concluded that only a slight influence was exerted by the reinforced soil's Young's modulus on the change of the tunnel diameter.

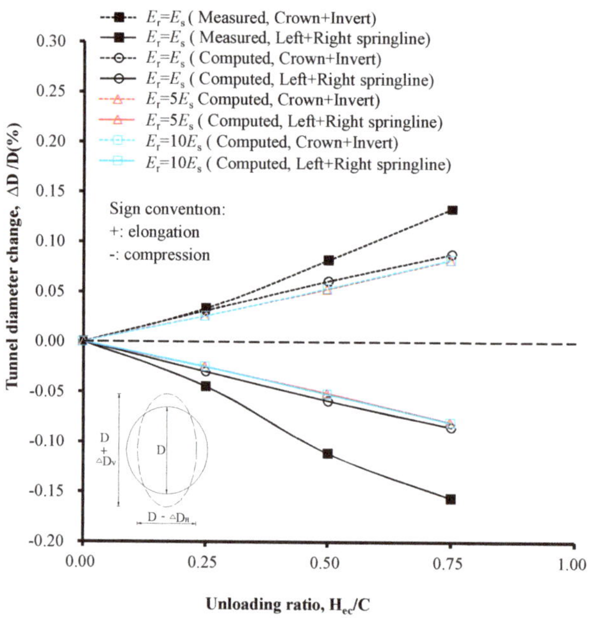


**Figure 7.** Measured and computed changes in tunnel diameter with different soil reinforcement moduli.

### 3.3. Effect of Reinforcement Depth on Tunnel Heave

Figure 8 compares the measured and computed normalized tunnel heave and maximum tunnel heave variation with different reinforcement depths, where $H_e$ is the final excavation depth. When none of the soil elements were reinforced, the computed maximum tunnel heave was 0.116% $H_e$. When the soil area S1 was reinforced, the computed tunnel heave decreased from 0.116% $H_e$ to 0.099% $H_e$ with a constant reinforcement Young's modulus. However, significant changes were not observed when the reinforcement depth increased further. Thus, it is obvious that the soil reinforcement depth underneath the basement reduced the tunnel heave, even for a thin layer. The further increase of the reinforcement depth did not have an effect and was uneconomical.

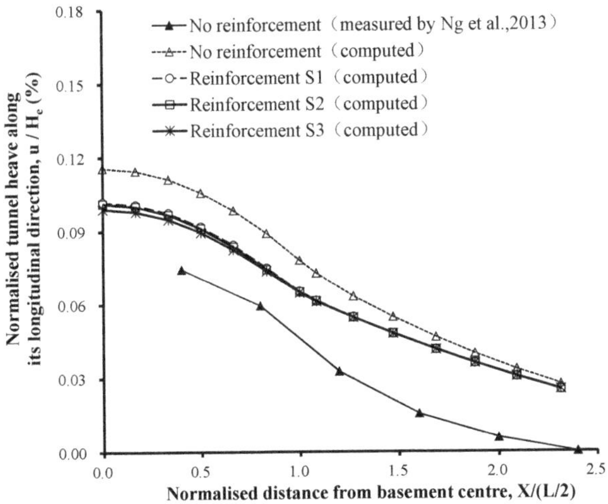


(**a**)

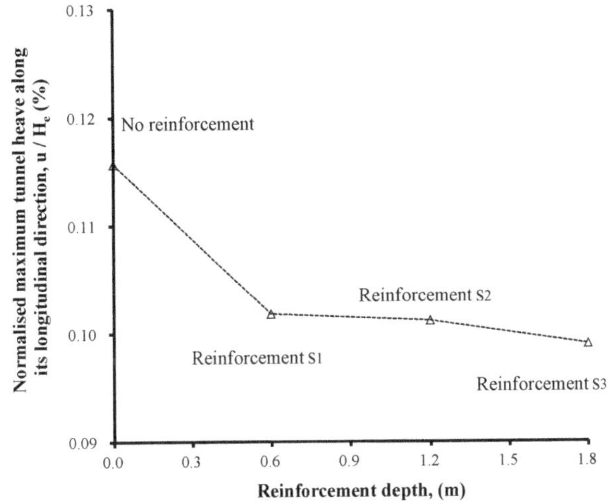


(**b**)

**Figure 8.** Measured and computed (**a**) tunnel heave and (**b**) maximum tunnel heave with different reinforced soil areas.

### 3.4. *Effect of Reinforcement Depth on Tunnel Diameter Change*

Figure 9 shows the measured and computed changes in the tunnel diameter with different soil reinforcement depths. The positive and negative values denote the elongation and compression of the tunnel, respectively. It is also noted that the described tunnel deformation was located in the tunnel section in a circumferential direction. Tunnel diameter was elongated in the vertical direction and compressed in the horizontal direction after stress relief. The elongated and compressed values increased with the unloading ratio. As the reinforcement depth increased, the elongation and compression of the tunnel lining decreased. Obvious changes were not observed when the reinforcement depth increased further. Thus, it was concluded that the reinforcement depth exerted only a slight influence on the change of the tunnel diameter. When the unloading ratio was 0.5 and the soil area S3 was reinforced, the tunnel diameter change was 0.061%. Compared with the 0.049%

change for no reinforcement, a maximum tunnel diameter change ratio of 19% occurred. A summary of the comparison between the computed and measured values is presented in Table 2.

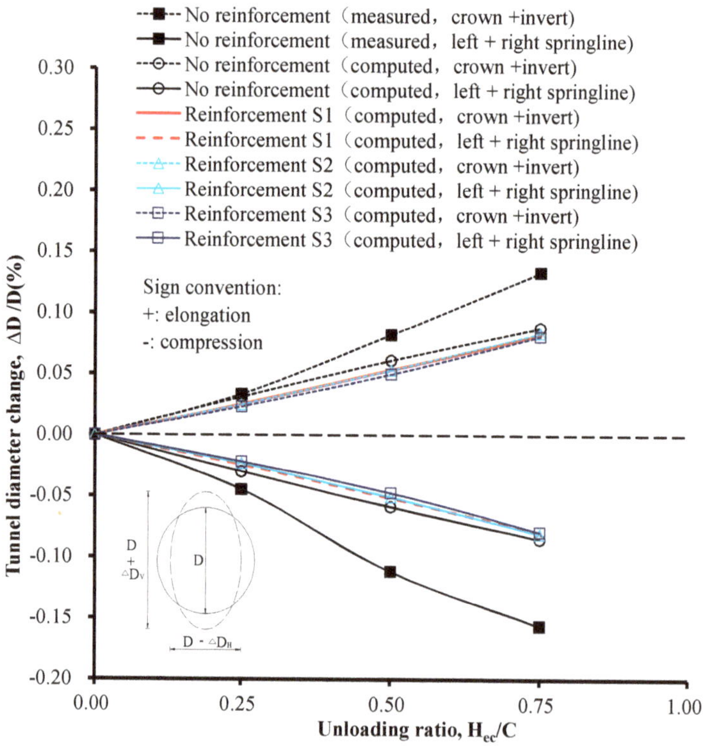


**Figure 9.** Measured and computed change in tunnel diameter with different soil reinforcement areas.

### *3.5. Analyses of Soil Responses Caused by Excavation*

#### 3.5.1. Stress Distributions of Soil Elements in the Tunnel's Longitudinal Direction

To better understand the tunnel behavior in the transverse and longitudinal directions, the stress distribution of the soil elements around the tunnel are presented in Figures 10 and 11. Figure 10a shows the computed changes in the vertical stress at the tunnel crown in the longitudinal direction for different reinforcement Young's moduli. The positive and negative values denote the increase and decrease in the stress acting on the tunnel lining, respectively. Along the tunnel crown, the vertical stress beneath the basement reduced significantly as a result of the excavation. An almost uniform stress relief was observed beneath the basement. The stress was concentrated beneath the diaphragm wall. This is because the tunnel moved upward due to stress relief, whereas the trend was blocked by the diaphragm wall. The concentrated stress may increase largely beneath the diaphragm wall and then dissipate behind the wall. Thus, it is possible that there will be an increase in stress, which has also been observed in the literature [14]. The vertical stress in the soil increased by more than 100 kPa. After the basement excavation, the maximum change in the vertical stress at the tunnel crown exceeded the allowable limit (i.e., ±20 kPa) set by the Building Department of Hong Kong [27]. In the case where the soil element was not reinforced, the maximum reduction in the vertical stress was 159 kPa. When the reinforced soil's Young's modulus increased from $E_s$ to $5E_s$, the maximum reduction in the vertical stress was 120 kPa. This was because the reinforced soil blocked the stress transfer and thus reduced the stress release around the tunnel. However, obvious stress changes were not observed when the reinforced soil's Young's modulus increased from $5E_s$ to $10E_s$, which can explain the phenomenon shown in Figure 6. Figure 10b shows the computed changes in the vertical stress at the tunnel crown in the longitudinal direction for different reinforcement depths. As can be seen, the maximum stress reduction was restrained from 159 kPa to 120 kPa when the reinforcement depth changed from 0 to 0.6 m. Further obvious changes were not observed for a reinforcement depth larger than 0.6 m, which is consistent with the tunnel heave change shown in Figure 8.

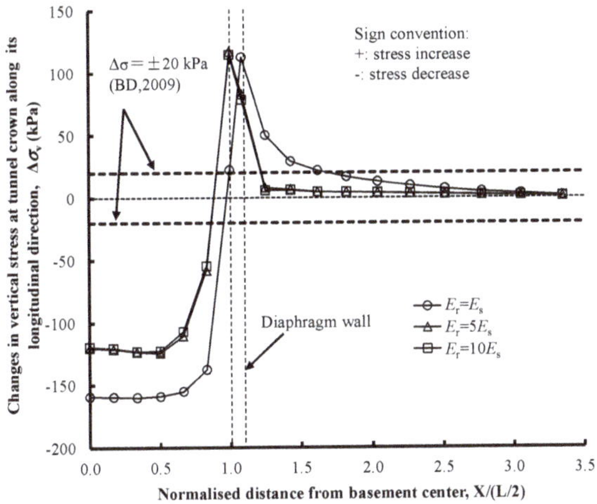


(**a**)

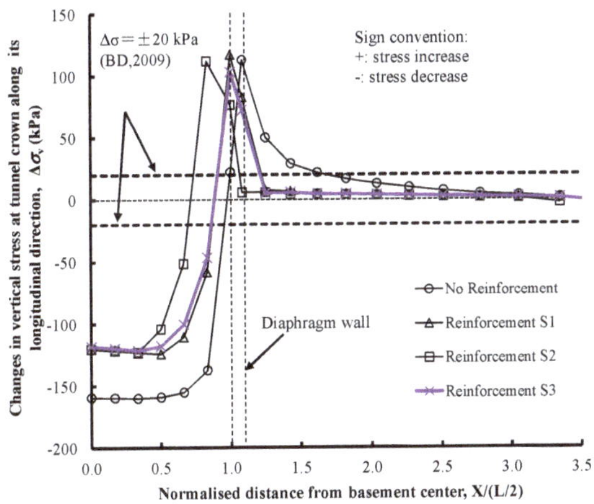


(**b**)

**Figure 10.** Computed vertical stress change in the soil element at the crown in the circumferential direction for (**a**) different reinforcement Young's moduli and (**b**) different reinforcement depth.

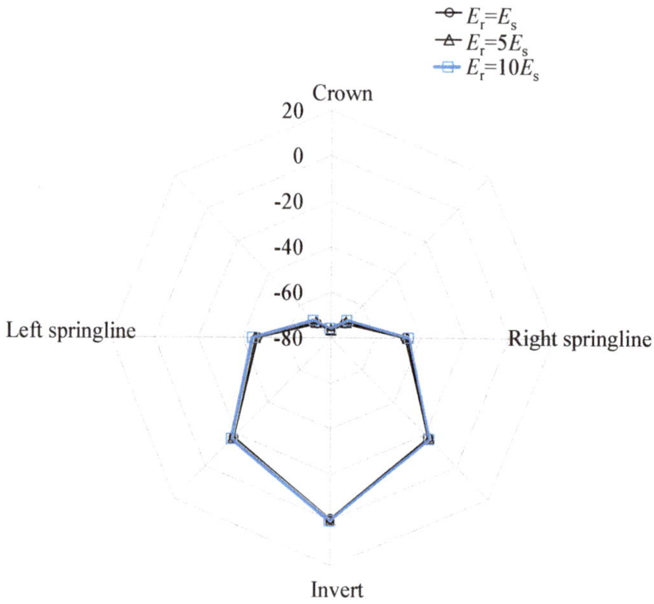


(a)

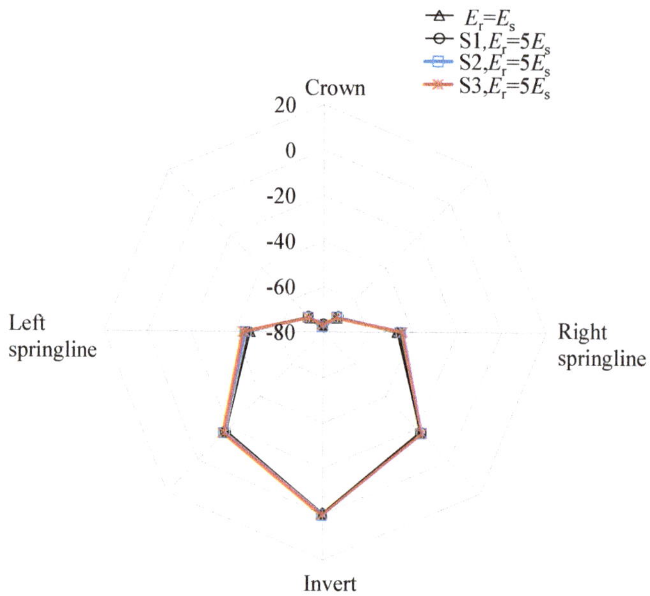


(b)

**Figure 11.** Computed earth pressure change in the soil element along the tunnel transverse direction for (**a**) different reinforcement Young's moduli and (**b**) different reinforcement depth.

### 3.5.2. Stress Distributions of Soil Elements in the Tunnel's Transverse Direction

Figure 11 shows the normal stress change distribution of the soil elements in the transverse direction of the tunnel which was located in the section under the center of the basement. As can be seen, the maximum and minimum stress changes were 80 kPa and almost 0 kPa, respectively, and occurred at the top of the tunnel and at the invert, respectively. It was found that the stress at the invert changed little. This may be due to the rather large stiffness of the tunnel which prevented the stress change of the soil at the tunnel invert. The stress change in the soil elements around the tunnel springlines was 45 kPa. Thus, the tunnel circumference was elongated in the vertical direction and compressed in the horizontal direction, as shown in Figures 7 and 9. As can be seen in Figure 10, in all reinforced cases, only a slight stress change could be observed for the soil elements around the tunnel, in comparison to the non-reinforced case. This can be explained by the diameter change, as shown in Figures 7 and 9.

## 4. Conclusions

This study evaluated the effect of the reinforced soil's Young's modulus and the effect of the reinforcement depth on the tunnel response as a result of a basement excavation, based on back-analyzing a centrifuge model test with no reinforcement. The tunnel heave and diameter change and the stress changes both in the longitudinal and in the transverse directions, were analyzed. The obtained results can provide reference for practical engineering in terms of applying construction measures to control tunnel deformation caused by stress relief. Based on the above analyses, the following conclusions were drawn:

(a) As a result of the stress relief caused by the basement excavation, tunnel heave was observed for the tunnel underneath the basement excavation. The heave was greatly reduced by improving the basement bottom soil's Young's modulus. When a specific soil area S1 was reinforced with a Young's modulus that varied from $E_s$ to $5E_s$, the computed tunnel heave decreased by 18%. This was because the reinforced soil blocked the stress transfer and thus reduced the tunnel heave caused by excavation unloading. A further increase of the reinforced soil's Young's modulus did not significantly reduce the tunnel heave. This can be attributed to the fact that obvious stress changes were not observed when the reinforced soil's Young's modulus increased from $5E_s$ to $10E_s$.

(b) When the reinforcement depth was 0.6 m, the computed tunnel heave decreased by 15% in comparison with the non-reinforced case. However, significant changes were not observed when the reinforcement depth increased further. The further increase of reinforcement depth had no effect and would therefore be uneconomical.

(c) As the reinforced soil's Young's modulus varied from $E_s$ to $5E_s$, the maximum elongation of the tunnel lining in the circumferential direction changed by 9%. Significant changes were not observed when the reinforced soil's Young's modulus increased from $5E_s$ to $10E_s$. Significant changes were not observed when the reinforcement depth increased further. Thus, it was concluded that the reinforced soil's Young's modulus and the reinforcement depth exerted only a slight influence on the diameter change of the tunnel.

**Author Contributions:** Conceptualization, H.S.; methodology, H.S.; software, H.S.; validation, W.S.; formal analysis, H.S. and W.S.; investigation, H.S. and W.S.; resources, W.S.; data curation, H.S.; writing—original draft preparation, H.S.; writing—review and editing, H.S. and W.S.; visualization, H.S.; supervision, H.S.; project administration, H.S.; funding acquisition, H.S.

**Funding:** This study was sponsored by the National Natural Science Foundation of China (grant number 51708245), the Natural Science Foundation of Jiangsu Province (grant number BK20160426), the Construction System Science and Technology Project of Jiangsu Province (grant number 2017ZD124), the Natural Science Foundation for Colleges and Universities in Jiangsu Province (17KJB130003, 17KJA560001), the Project with Science and Technology of Huai'an City (HAG201606).

**Acknowledgments:** We thank Liwen Bianji, Edanz Editing China (www.liwenbianji.cn/ac) for editing the English text of a draft of this manuscript.

**Conflicts of Interest:** The authors declare that there is no conflict of interest regarding the publication of this paper.

**Data Availability Statement:** The computed data used to support the findings of this study are included within the article. Previously reported measured data were used to support this study and are available at [https://doi.org/10.1139/cgj-2012-0423, https://doi.org/10.1139/cgj-2014-0361]. These prior studies (and datasets) are cited at relevant places within the text as references [4,14].

## References

1. Burford, D. Heave of tunnels beneath the Shell Centre. London, 1959–1986. *Géotechnique* **1988**, *38*, 135–137. [CrossRef]
2. Chang, C.-T.; Sun, C.-W.; Duann, S.; Hwang, R.N. Response of a Taipei Rapid Transit System (TRTS) tunnel to adjacent excavation. *Tunn. Undergr. Space Technol.* **2001**, *16*, 151–158. [CrossRef]
3. Simpson, B.; Vardanega, P.J. Results of monitoring at the British Library excavation. *Proc. Inst. Civ. Eng.—Geotech. Eng.* **2014**, *167*, 99–116. [CrossRef]
4. Ng, C.W.; Shi, J.; Hong, Y. Three-dimensional centrifuge modelling of basement excavation effects on an existing tunnel in dry sand. *Can. Geotech. J.* **2013**, *50*, 874–888. [CrossRef]
5. Huang, H.; Huang, X.; Zhang, D. Centrifuge modelling of deep excavation over existing tunnels. *Proc. Inst. Civ. Eng.—Geotech. Eng.* **2014**, *167*, 3–18. [CrossRef]
6. Ng, C.; Shi, J.; Lei, G.; Mašín, D.; Sun, H. Influence of sand density and retaining wall stiffness on three-dimensional responses of tunnel to basement excavation. *Can. Geotech. J.* **2015**, *52*, 1811–1829. [CrossRef]
7. Lo, K.; Ramsay, J. The effect of construction on existing subway tunnels—A case study from Toronto. *Tunn. Undergr. Space Technol.* **1991**, *6*, 287–297. [CrossRef]
8. Doležalová, M. Tunnel complex unloaded by a deep excavation. *Comput. Geotech.* **2001**, *28*, 469–493. [CrossRef]
9. Sharma, J.S.; Hefny, A.M.; Zhao, J.; Chan, C.W. Effect of large excavation on displacement of adjacent MRT tunnels. *Tunn. Undergr. Space Technol.* **2001**, *16*, 93–98. [CrossRef]
10. Hu, Z.F.; Yue, Z.Q.; Zhou, J.; Tham, L.G. Design and construction of a deep excavation in soft clay adjacent to the shanghai metro tunnels. *Can. Geotech. J.* **2003**, *40*, 933–948. [CrossRef]
11. Zheng, G.; Wei, S.W. Numerical analysis of influence of overlying pit excavation on existing tunnels. *J. Cent. South Univ. Technol.* **2008**, *15*, 69–75. [CrossRef]
12. Liu, H.L.; Li, P.; Liu, J.Y. Numerical investigation of underlying tunnel heave during a new tunnel construction. *Tunn. Undergr. Space Technol.* **2011**, *26*, 276–283. [CrossRef]
13. Huang, X.; Schweiger, H.F.; Huang, H.W. Influence of deep excavations on nearby existing tunnels. *Int. J. Geomech. ASCE* **2013**, *13*, 170–180. [CrossRef]
14. Ng, C.W.W.; Sun, H.S.; Lei, G.H.; Shi, J.W.; Mašín, D. Ability of three different soil constitutive models to predict a tunnel's response to basement excavation. *Can. Geotech. J.* **2015**, *52*, 1685–1698. [CrossRef]
15. Shi, J.; Ng, C.W.W.; Chen, Y. Three-dimensional numerical parametric study of the influence of basement excavation on existing tunnel. *Comput. Geotech.* **2015**, *63*, 146–158. [CrossRef]
16. Zhang, J.F.; Chen, J.J.; Wang, J.H.; Zhu, Y.F. Prediction of tunnel displacement induced by adjacent excavation in soft soil. *Tunn. Undergr. Space Technol.* **2013**, *36*, 24–33. [CrossRef]
17. Zhang, Z.; Huang, M.; Wang, W. Evaluation of deformation response for adjacent tunnels due to soil unloading in excavation engineering. *Tunn. Undergr. Space Technol.* **2013**, *38*, 244–253. [CrossRef]
18. Zhang, Z.; Zhang, M.; Zhao, Q. A simplified analysis for deformation behavior of buried pipelines considering disturbance effects of underground excavation in soft clays. *Arab. J. Geosci.* **2015**, *8*, 1–15. [CrossRef]
19. Liang, R.; Wu, W.; Yu, F. Simplified method for evaluating shield tunnel deformation due to adjacent excavation. *Tunn. Undergr. Space Technol.* **2018**, *71*, 94–105. [CrossRef]
20. Zheng, G.; Du, Y.M.; Diao, Y. Influenced zones for deformation of existing tunnels adjacent to excavations. *Chin. J. Geotech. Eng.* **2016**, *38*, 599–612.
21. Ng, C.W.W.; Van Laak, P.A.; Tang, W.H.; Li, X.S.; Zhang, L.M. The Hong Kong geotechnical centrifuge. In Proceedings of the 3rd International Conference on Soft Soil Engineering, Hong Kong, China, 6–8 December 2001; pp. 225–230.

22. Ng, C.W.W.; Van Laak, P.A.; Zhang, L.M.; Tang, W.H.; Zong, G.H.; Wang, Z.L.; Xu, G.M.; Liu, S.H. Development of a four-axis robotic manipulator for centrifuge modeling at HKUST. In Proceedings of the International Conference on Physical Modelling in Geotechnics, St John's, NL, Canada, 10–12 July 2002; pp. 71–76.
23. Ng, C.W.W. The state-of-the-art centrifuge modelling of geotechnical problems at hkust. *J. Zhejiang Univ. A* **2014**, *15*, 1–21. [CrossRef]
24. ABAQUS, Inc. *ABAQUS User's and Theory Manuals*; Version 6.17; ABAQUS, Inc.: Providence Rhode Island, RI, USA, 2017.
25. LTA. *Code of Practice for Railway Protection*; Development & Building Control Department, Land Transport Authority (LTA): Singapore, 2000.
26. BTS. *Specification for Tunnelling*; British Tunnelling Society (BTS): Thomas Telford, London, UK, 2000.
27. BD. *Practice Note for Authorized Persons APP-24. Technical Notes for Guidance in Assessing the Effects of Civil Engineering Construction/Building Development on Railway Structures and Operations*; Building Department of the Government of HKSAR (BD): Hongkong, China, 2009.

applied sciences

MDPI

Article

# Quantitative Analysis of Soil Displacement Induced by Ground Loss and Shield Machine Mechanical Effect in Metro Tunnel Construction

**Yimo Zhu [1], Liang Chen [2,*], Heng Zhang [1,*], Peng Tu [3] and Shougen Chen [1]**

1 Key Laboratory of Transportation Tunnel Engineering, Ministry of Education, Southwest Jiaotong University, Chengdu 610031, China
2 School of Civil Engineering, Changsha University of Science and Technology, Changsha 410076, China
3 Sichuan Vocational and Technical College of Communications, Chengdu 611130, China
* Correspondence: liangchen_2018@163.com (L.C.); Hengzhang1985@outlook.com (H.Z.)

Received: 26 June 2019; Accepted: 24 July 2019; Published: 26 July 2019

**Abstract:** In order to relieve the increasing ground traffic pressure in the process of urbanization in China, it is inevitable to build more metro lines. However, the stratum movement caused by tunneling affects the safety of adjacent underground structures and aboveground buildings. Therefore, how to evaluate and control the stratum movement is a prominent problem. In this paper, based on the engineering project of an interval tunnel between Shizishan Station and Chuanshi Station in Chengdu Metro Line 7, China, the action mechanism of stratum movement induced by shield tunneling is analyzed, and the effect factors are divided into two categories: ground loss factors and mechanical factors. Combining the advantages of Loganathan method and mirror source-sink method, a new solution of three-dimensional displacement induced by ground loss is proposed. Based on the elastic half-space Mindlin model, the displacement at any point induced by four mechanical effect factors is deduced. Finally, the total displacement is verified by field monitoring data and quantitative analyzed in various parts.

**Keywords:** metro constructions; shield tunnel; ground settlement; soil displacement; analytical; Mindlin solution

## 1. Introduction

With the acceleration of urbanization in China, the growth rate of urban land use is much lower than the urban population growth rate. The limited number of urban surface land resources can no longer meet people's needs for living space. The development direction of urban space is gradually changing from the horizontal direction of extension to the vertical direction of intension, which has resulted in many underground structures [1–4]. Although the development of urban underground space has detailed plans and arrangements before construction, it still has some unpredictability in the long term. Therefore, the negative conditions that new underground structures create near the existing ones happen from time to time. As the main artery of underground traffic in modern cities, this phenomenon of metro tunnels is particularly prominent. Among many construction methods of metro tunnel, shield tunneling is becoming the preferred construction scheme because of its advantages of fast construction and small disturbance to stratum [5–9]. Although shield tunneling has many advantages as mentioned above, and the construction technology has made great progress in many years, due to the defects of geological conditions and construction technology, the advance of shield tunneling will inevitably create a disturbance, change the stress state of soil and cause stratum displacement. Therefore, how to evaluate and control the stratum displacement has always been a concern for engineers [10–12].

The important of this problem has prompted researchers to study many methods for predicting stratum displacement induced by shield tunneling. Based on the factors affecting stratum displacement, the methods can be divided into two categories. The first category concerns the ground loss factors during shield excavation, and the second addresses the mechanical effect factors. In terms of ground loss factors, empirical solution is one of the most commonly used methods. Peck assumed that the curve of land settlement trough satisfies the conditions of Gauss distribution and invariant stratum volume, as was deduced Peck's empirical formula based on field monitoring data [13]. Celestino et al. [14] and many other researchers [15–19] noted that Peck's empirical formula cannot accurately describe soil settlement trough in many cases and proposed improved empirical formulas under different geologic conditions. Besides empirical solution, there are also many researchers who contribute significantly through analytical solution [20–23], numerical simulation [24–26] and model testing [27,28]. In terms of mechanical effect factors, Mindlin's solution is one of the well-known solutions to calculate the stress and displacement caused by a point load at an embedment depth in an elastic half-space [29]. Based on Mindlin's solution, many modified analytical solutions are proposed [30–33]. For instance, Verruijt et al. derived a theoretical solution for the deformations of an elastic half-plane with a circular cavity, which is often used in tunnel excavation. [34,35].

However, most previous studies only consider the stratum displacement caused by ground loss factors, ignoring the influence of mechanical factors without clear theoretical basis [7,8,17,21]. A few studies containing mechanical factors are not thorough enough, and they are usually shown in the form of total displacement without the quantitative analysis of the proportion of stratum displacement caused by each factor [6,9,10]. In this paper, based on the engineering project of an interval tunnel between Shizishan Station and Chuanshi Station in Chengdu Metro Line 7, China, combining Loganathan empirical formula and mirror source-sink method, the stratum displacement caused by ground loss factors is calculated. The mechanical effect factors are divided into four parts and calculated based on the Mindlin solution. The proportion of stratum displacement caused by each factor is quantitatively analyzed, which can provide guidance for the stratum displacement calculation of shield tunneling in the future.

## 2. Engineering Background

The engineering project in this paper is an interval tunnel between Shizishan Station and Chuanshi Station in Chengdu Metro Line 7, China. The tunnel consists of two single-track tunnels with a total length of 970.4 m and a burial depth of 15 m, which is mainly located in weathered mudstone stratum and excavated by EPB (Earth Pressure Balanced) shield. The external diameter of shield and fabricated precast segment are 6 m and 6.28 m, respectively. The diagram of shield tunneling in diagram is shown in Figure 1.

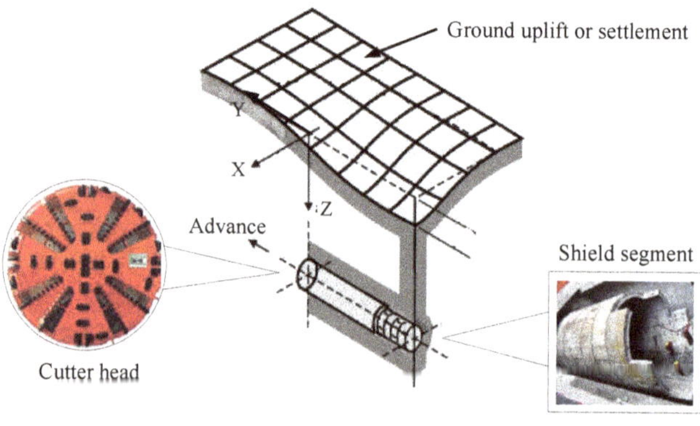


**Figure 1.** The diagram of shield tunneling in stratum.

In order to carry out the calculations, the parameters that need to be determined can be divided into two categories. The first category concerns the physical parameters of soil, such as elastic modules, Poisson's ratio, etc. They are obtained from engineering geological reports and field experiments, as shown in Table 1. The second is the construction parameters of the shield machine, such as shield tunneling speed, strain rate of grouting layer, etc. The equipment used in this project is CREC153 EPB shield machine produced by China Railway Engineering Equipment Group, its parameters are shown in Table 2.

**Table 1.** Physical parameters of soil.

| Material | Unit Weight ($kN/m^3$) | Thickness (m) | Poisson's Ratio | Cohesion (kPa) | Friction Angle | Elastic Modules (MPa) |
|---|---|---|---|---|---|---|
| Topsoil | 20 | 2.1 | 0.3 | 45 | 14° | 5.94 |
| Intermediary weathered mudstone | 23.4 | 37.9 | 0.17 | 300 | 18° | 180 |

**Table 2.** Construction parameters of shield machine.

| Parameter | Shield Tunneling Speed (mm/min) | Rotation Speed of Cutter Head (r/min) | Cutter Opening Ratio (%) | Closure Number of Cutters | Diameter of Shield (mm) |
|---|---|---|---|---|---|
| Value | 80 | 3.55 | 36 | 4 | 6280 |
| **Length of Main Machine (mm)** | **Diameter of Segment (mm)** | **Synchronous Grouting Pressure (MPa)** | **Strain Rate of Grouting Layer (%)** | **Slip Softening Coefficient** | **Interface Friction Angle (°)** |
| 8427 | 6000 | 0.3 | 8.4 | 0.9 | 6.5 |

## 3. Ground Loss Effect Factors

Ground loss $g$ is the primary cause of stratum settlement and deformation. During shield tunnel construction, the causes of ground loss include curved propulsion, shield steering adjustment, soil thrusting into the shield tail gap, head-down propulsion, head-up propulsion, and tunnel lining deformation. The ground loss can be calculated by using the Lee's equation [36].

$$g = G_P + U_p + \omega \tag{1}$$

where $G_p$ is physical gap, it is the gap between the outer wall of the shield shell and the outer wall of the segment. In the field, synchronous grouting is used to decrease the physical gap, but due to the hydration reaction characteristics and the water permeation loss of cement slurry, there is still a residual gap $G_p'$ when the physical gap $G_p$ is fully filled by grouting. Usually, $G_p' = G_p \times \beta_l$, $\beta_l$ is slurry strain rate; $U_p$ is the plastic deformation of shield working face during tunnel excavation. Generally, $U_p = 0$ when the EPB shield is used; $\omega$ is the additional ground loss induced by shield rectification, head-up propulsion, head-down propulsion and the soil consolidation in the tunneling disturbance zone. $\omega$ can be assumed to be 0 because the working face thrust of shield is equal to the soil pressure during most of the time.

### 3.1. Displacement Solution

Based on the Verruijt analytical formula [34,35], considering quantify ground loss with non-equivalent radial displacement, Loganathan [21] put forward a method for estimating stratum displacement caused by ground loss in undrained clays; the formula is as shown in Equation (2).

$$\begin{cases} S_z = R^2\left\{-\frac{z-H}{x^2+(z-H)^2} + (3-4\mu)\frac{z+H}{x^2+(z+H)^2} - \frac{2z\left[x^2-(z+H)^2\right]}{\left[x^2+(z+H)^2\right]^2}\right\} \times \varepsilon_{x,z} \\ S_x = -R^2 x\left\{\frac{1}{x^2+(z-H)^2} + \frac{(3-4\mu_u)}{x^2+(z+H)^2} - \frac{4z(z+H)}{\left[x^2+(z+H)^2\right]^2}\right\} \times \varepsilon_{x,z} \end{cases} \tag{2}$$

where $R$ is tunnel external radius; $H$ is burial depth of tunnel center; $\mu$ is Poisson's ratio of soil; $\varepsilon_{x,z}$ is the parameter of non-equivalent radial displacement, fitting by exponential function based on measured data, which can be expressed as Equation (3).

$$\varepsilon_{x,z} = \frac{4gR + g^2}{4R^2}\exp\left\{-\left|\frac{1.38x^2}{(H+R)^2} + \frac{0.69z^2}{H^2}\right|\right\} \tag{3}$$

Considering that the displacement solution caused by ground loss derived by Loganathan is a two-dimensional solution, which is often a three-dimensional problem in actual engineering, it is necessary to extend the two-dimensional displacement solution to three-dimensional. The three-dimensional ground loss parameter $\varepsilon_{x,z,y}$ can be extracted from the Sagaseta' solution [20].

Sagaseta deduced the expression of ground displacement caused by tunnel construction by using the mirror source-sink method.

$$\begin{cases} S_{x0} = -\frac{v_{loss}}{2\pi}\frac{x}{x^2+H^2}\left[1 + \frac{y}{(x^2+y^2+H^2)^{\frac{1}{2}}}\right] \\ S_{y0} = \frac{v_{loss}}{2\pi}\frac{1}{(x^2+y^2+H^2)^{\frac{1}{2}}} \\ S_{z0} = \frac{v_{loss}}{2\pi}\frac{H}{x^2+H^2}\left[1 + \frac{y}{(x^2+y^2+H^2)^{\frac{1}{2}}}\right] \\ S_{x0}(y \to \infty) = -\frac{v_{loss}}{\pi}\frac{x}{x^2+H^2} \\ S_{z0}(y \to \infty) = \frac{v_{loss}}{\pi}\frac{H}{x^2+H^2} \end{cases} \tag{4}$$

where $V_{loss}$ is ground loss. From Equation (4) it can be found:

$$\begin{cases} S_{x0}(y) = S_{x0}(y \to \infty) \times \frac{1}{2} \times \left[1 + \frac{y}{(x^2+y^2+H^2)^{\frac{1}{2}}}\right] \\ S_{z0}(y) = S_{z0}(y \to \infty) \times \frac{1}{2} \times \left[1 + \frac{y}{(x^2+y^2+H^2)^{\frac{1}{2}}}\right] \end{cases} \tag{5}$$

Equation (5) shows that the distribution of stratum displacement caused by shield construction (ground loss) along the Y axis satisfies the law of $\frac{1}{2} \times [1 + y/(x^2 + y^2 + z^2)^{\frac{1}{2}}]$. Therefore, based on this law, the Loganathan's solution is extended to the three-dimensional. It is noted that the coordinate system adopted in this paper is different from that used in the Sagaseta's solution, so the distribution law should be revised, and its expression is as Equation (6):

$$\varepsilon_y = \frac{1}{2} \times \left[1 - \frac{y}{(x^2+y^2+H^2)^{\frac{1}{2}}}\right] \tag{6}$$

The plane ground loss parameter $\varepsilon_{x,z}$ is extended to three-dimensional parameter $\varepsilon_{x,z,y}$ which can be seen as Equation (7). Then, three-dimensional stratum displacement solution, Equation (8) can be obtained by replacing $\varepsilon_{x,z}$ in Equation (2).

$$\varepsilon_{x,y,x} = \varepsilon_y \times \varepsilon_{x,z} = \left[1 - \frac{y}{(x^2+y^2+H^2)^{\frac{1}{2}}}\right] \times \frac{4gR + g^2}{8R^2}\exp\left\{-\left|\frac{1.38x^2}{(H+R)^2} + \frac{0.69z^2}{H^2}\right|\right\} \tag{7}$$

$$\begin{cases} S_z = R^2\left\{-\frac{z-H}{x^2+(z-H)^2} + (3-4\mu)\frac{z+H}{x^2+(z+H)^2} - \frac{2z\left[x^2-(z+H)^2\right]}{\left[x^2+(z+H)^2\right]^2}\right\} \times \varepsilon_{x,y,z} \\ S_x = -R^2x\left\{\frac{1}{x^2+(z-H)^2} + \frac{(3-4\mu)}{x^2+(z+H)^2} - \frac{4z(z+H)}{\left[x^2+(z+H)^2\right]^2}\right\} \times \varepsilon_{x,y,z} \end{cases} \tag{8}$$

### 3.2. Quantitative Analysis

Ground loss $g$ and several parameters of Table 1 can be brought into Equation (8) to calculate ground loss displacement, and the counter map of ground settlement are shown in Figure 2. As we can see it, ground settlement at 4.5 D (D for diameter of tunnel) behind shield tail is stable to maximum 10.51 mm, and its value at 4.5 D in front of shield tail is only 10% of maximum. Therefore, it can be considered that the influence range of ground loss is within 4.5 D behind and in front of shield tail. As for deep soil layer, when the vertical distance from tunnel vault is greater than 0.5 D, the settlement has little changes. However, while it is less than 0.5 D, the settlement increases rapidly, as shown in Figure 3a. This is because that with the redistribution of stress, a pressure arch will be formed around the tunnel. The soil in the arch is a loosening zone, which has a greater displacement, and the soil out of the arch tends to be stable. As we can see from Figure 3b, horizontal displacement of deep soil is mainly distributed in 1 D from the bottom of the tunnel.

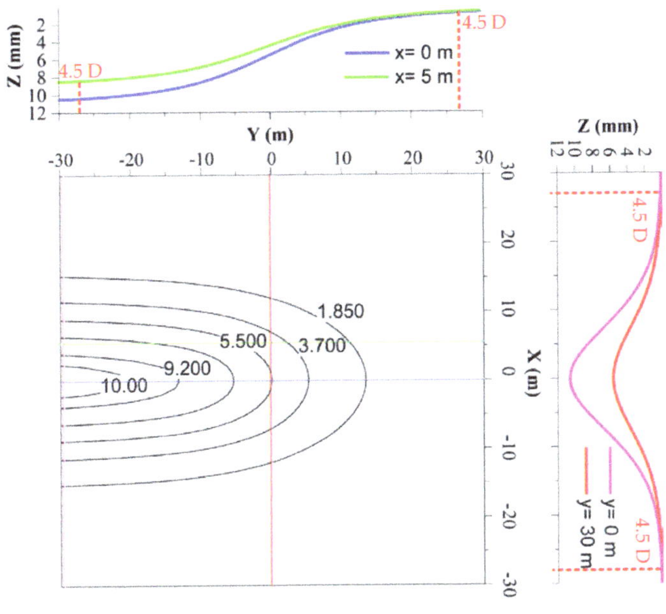


**Figure 2.** Counter map of ground settlement.

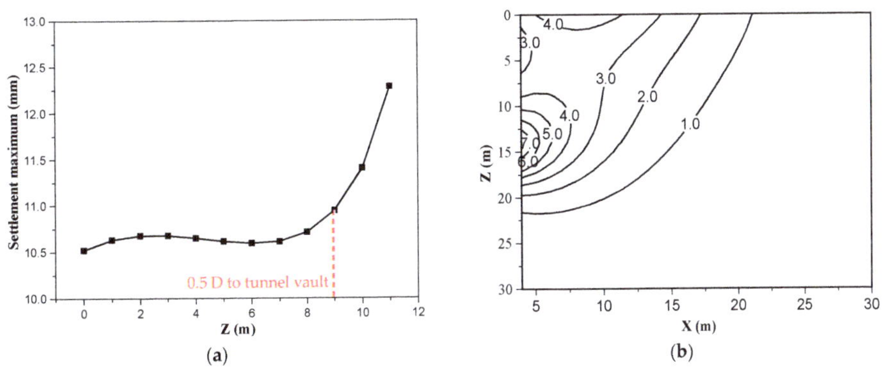


**Figure 3.** Displacement of stratum: (**a**) Variation curve of settlement maximum; (**b**) Counter map of horizontal displacement.

## 4. Mechanical Effect Factors

The shield machine has a complex mechanical interaction with the surrounding stratum when shield tunnel constructs. Based on the difference of sources and action modes, these mechanical effect factors can be divided into following categories, as Figure 4: Friction $f_d$ between the shield working face and the soil; Additional frontal thrust $q$ on the shield working face; Shell Friction $f_s$ between the shield and the soil; Radial pressure $p$ generated by synchronous grouting in shield tail gap.

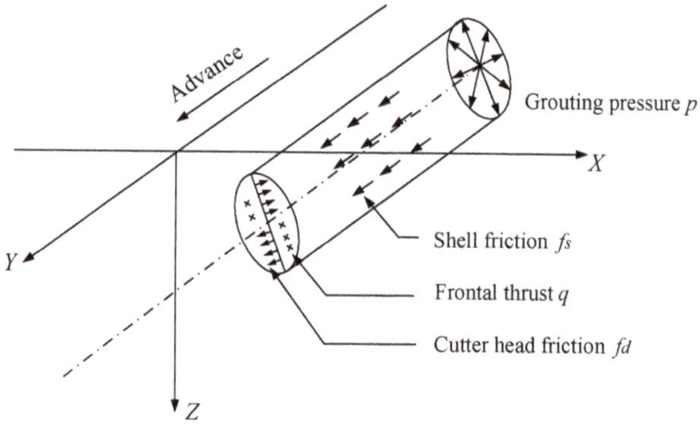


**Figure 4.** Four mechanical effect factors.

### 4.1. Mindlin's Solutions

The displacement caused by point load at any point can be solved by using Mindlin theoretical solution in the elastic half-space shown in Figure 5 [25]. The above four categories of force can be decomposed into vertical and horizontal forces, then brought into Mindlin solution to calculate.

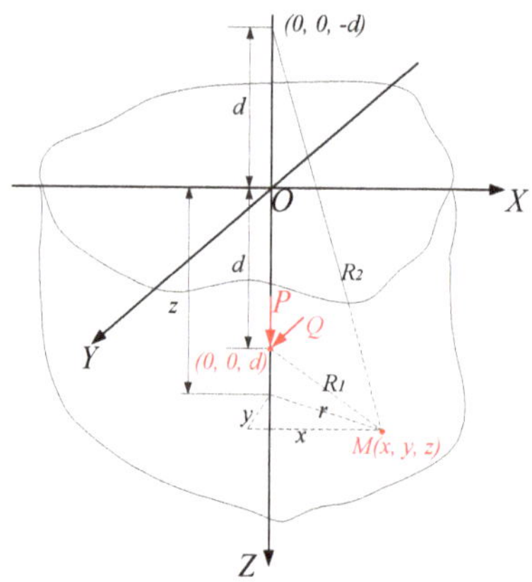


**Figure 5.** Mindlin model. Where $M$ $(x, y, z)$ is any point in the infinite half space; $P$ is a vertical concentrated load; $Q$ is a horizontal concentrated load; $d$ is the burial depth of concentrated load; $R_1 = \sqrt{r^2 + (z-d)^2}$; $R_2 = \sqrt{r^2 + (z+d)^2}$; $r = \sqrt{x^2 + y^2}$.

When there is a vertical concentrated load $P$ at burial depth $d$, then the displacement $S_{x(y),V}$ of x(y) direction and the displacement $S_{z,V}$ of z direction at any point $M$ $(x, y, z)$ in space induced by the load $P$ can be expressed as Equation (9):

$$\begin{cases} Sx(y), V = \frac{Px(y)}{16\pi G(1-\mu)}\left[\frac{z-d}{R_1^3} + \frac{(3-4\mu)(z-d)}{R_2^3} - \frac{4(1-\mu)(1-2\mu)}{R_2(R_2+z+d)} + \frac{6dz(z+d)}{R_2^5}\right] \\ Sz, V = \frac{P}{16\pi G(1-\mu)}\left[\frac{3-4\mu}{R_1} + \frac{8(1-\mu)^2-(3-4\mu)}{R_2} + \frac{(z-d)^2}{R_1^3} + \frac{(3-4\mu)(z+d)^2-2dz}{R_2^3} + \frac{6dz(z+d)^2}{R_2^5}\right] \end{cases} \tag{9}$$

where $\mu$ is Poisson's ratio; $G$ is soil shear modulus.

When there is a horizontal concentrated load $Q$ parallel to Y axis, pointing in the positive direction of the Y axis, then the displacement $S_{x,H}$ of x direction, $S_{y,H}$ of y, $S_{z,H}$ of z at any point $M$ $(x, y, z)$ in space induced by the load $Q$ can be expressed as Equation (10). While the horizontal concentrated load $Q$ parallel to X axis, due to the symmetry of x and y, it only needs to exchange x and y in Equation (10).

$$\begin{cases} S_{x,H} = \frac{Pxy}{16\pi G(1-\mu)}\left[\frac{1}{R_1^3} + \frac{3-4\mu}{R_2^3} - \frac{6dz}{R_2^5} - \frac{4(1-\mu)(1-2\mu)}{R_2(R_2+z+d)^2}\right. \\ S_{y,H} = \frac{P}{16\pi G(1-\mu)}\left[\frac{3-4\mu}{R_1} + \frac{1}{R_2} + \frac{y^2}{R_1^3} + \frac{(3-4\mu)y^2}{R_2^3} + \frac{2dz}{R_2^3}\left(1 - \frac{3y^2}{R_2^3}\right) + \frac{4(1-\mu)(1-2\mu)}{R_2+z+d}\left(1 - \frac{y^2}{R_2(R_2+z+d)}\right)\right] \\ S_{z,H} = \frac{Py}{16\pi G(1-\mu)}\left[\frac{z-d}{R_1^3} + \frac{(3-4\mu)(z-d)}{R_2^3} + \frac{4(1-\mu)(1-2\mu)}{R_2(R_2+z+d)} - \frac{6dz(z+d)}{R_2^5}\right] \end{cases} \tag{10}$$

Due to the location of load in Mindlin solution is at z axis, the coordinate of the formula above should be transformed to use when the location of load is at any point. The global coordinate system is xyz, let the local coordinate system be uvw and coordinate origin be $O'$ $(m, n, k)$, shown in Figure 6. When coordinate is transformed, xyz are replaced by uvw, then let $u = x - m$, $v = y - n$, $w = z - k$.

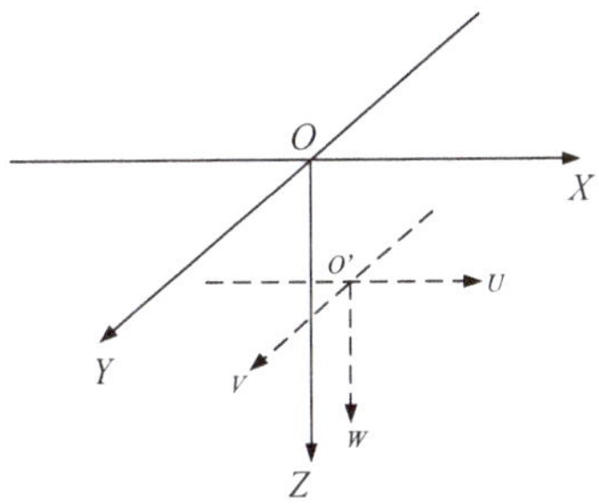


**Figure 6.** Coordinate transformation.

### 4.2. Quantitative Analysis of Additional Thrust q

When the shield machine advances, the whole machine is close to equilibrium state. However, large additional stresses still occur in the front of the cutter head, especially while plate-type panel cutter head with a small opening ratio is adopted. Wang built an extrusion model of cutter head and deduced the expression of additional thrust $q$ based on the field monitoring data, as shown in Equation (11) [37].

$$q = \frac{10.13(1-\mu)E_u\pi v(1-\xi)^2}{(1+\mu)(3-4\mu)Dk\omega} + \Delta p' \tag{11}$$

where $E_u$ is undrained elastic modulus; $v$ is shield tunneling speed; $\xi$ is cutter opening ratio; $D$ is diameter of shield; $k$ is closure number of cutters; $\omega$ is angular velocity of cutter head; $\Delta p'$ is extrusion pressure of cutters, generally takes 10–25 kPa.

Based on Equation (11) and parameters of Table 2, the additional thrust $q$ is 280 kPa. This is similar to the measured value in previous studies, which is much larger than the 20–30 kPa generally considered. The additional thrust $q$ is unit surface load parallel to Y axis. Based on the integral region in Figure 7 and Equation (10), let $m = \rho cos\ \theta$, $n = 0$, $k = 0$, depth of action of load element $d = H - \rho sin\ \theta$, stratum displacement induced by additional thrust $q$ can be obtained.

$$\begin{cases} S_{x,H}^{y,q} = \int_0^R \int_0^{2\pi} q\rho d\rho d\theta \frac{(x-\rho\cos\theta)y}{16\pi G(1-\mu)} \left[ \begin{array}{c} \frac{1}{R_1^3} + \frac{3-4\mu}{R_2^3} - \frac{6(H-\rho\sin\theta)z}{R_2^5} \\ -\frac{4(1-\mu)(1-2\mu)}{R_2(R_2+z+(H-\rho\sin\theta))^2} \end{array} \right] \\ S_{z,H}^{y,q} = \int_0^R \int_0^{2\pi} q\rho d\rho d\theta \frac{y}{16\pi G(1-\mu)} \left[ \begin{array}{c} \frac{z-(H-\rho\sin\theta)}{R_1^3} + \frac{(3-4\mu)(z-H+\rho\sin\theta)}{R_2^3} \\ +\frac{4(1-\mu)(1-2\mu)}{R_2(R_2+z+H-\rho\sin\theta)} \\ -\frac{6(H-\rho\sin\theta)z(z+H-\rho\sin\theta)}{R_2^5} \end{array} \right] \end{cases} \tag{12}$$

where $R_1 = \sqrt{r^2 + (z - H + \rho\sin\theta)^2}$; $R_2 = \sqrt{r^2 + (z + H - \rho\sin\theta)^2}$; $r = \sqrt{(x - \rho\cos\theta)^2 + y^2}$; $H$ is burial depth of tunnel center; $\mu$ is Poisson's ratio of soil; $\rho$ is the distance to the center of cutter head.

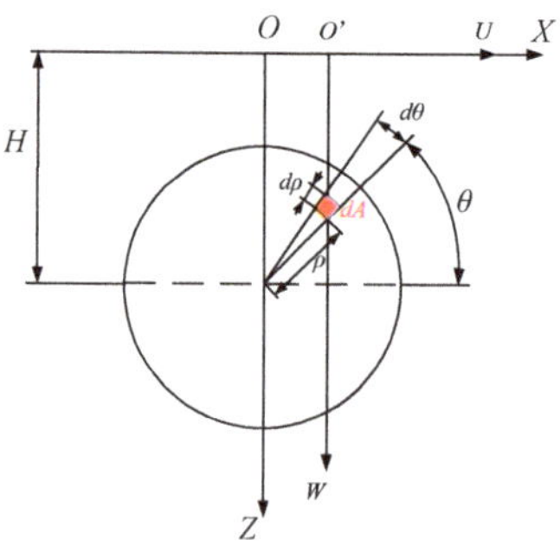


**Figure 7.** Integral region of $q$.

As shown in Figure 8, under the action of additional thrust $q$, ground uplifts in front of the cutter head, settlements behind the cutter head. Both reach maximum at 1 D and tend to be stable at 4.5 D away from the cutter head. The displacement is small by and large, for each 50 kPa increase in $q$, the maximum displacement increases by 0.025 mm. Figure 9 shows that the horizontal displacement caused by $q$ of the cutter head mainly appears within the depth of 1.5 D when Y = 1.

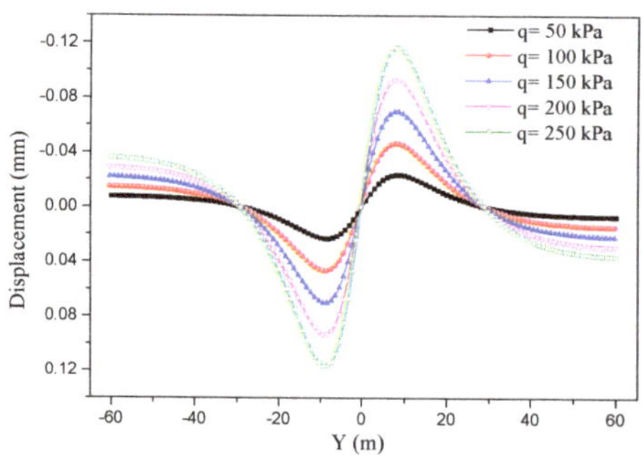


**Figure 8.** Longitudinal ground settlement caused by $q$.

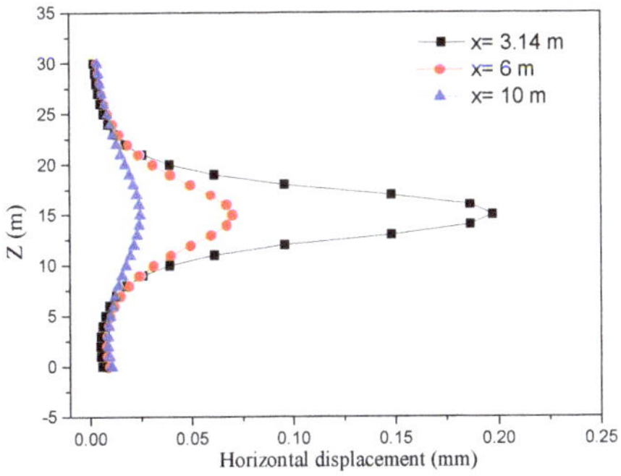


**Figure 9.** Horizontal displacement of stratum caused by $q$.

## 4.3. *Quantitative Analysis of Frictional Force* $f_d$

Based on the integral region shown in Figure 7, frictional force $f_d$ between the shield working face and the soil is expressed as Equation (12), which can be decomposed into horizontal load $f_{dH}$ and vertical load $f_{dV}$, as Equation (13). The solution of total stratum displacement can be obtained by compositing the displacements of Equation (10).

$$f_d = [\gamma(H - \rho \sin\theta)\frac{\mu}{1-\mu} + q] \times \varphi_f \tag{13}$$

$$\begin{cases} f_{dH} = f_d \sin\theta \\ f_{dV} = f_d \cos\theta \end{cases} \tag{14}$$

$$\begin{cases} S_x^{f_d} = S_{x,H}^{x,f_{dH}} + S_{x,V}^{f_{dV}} \\ S_z^{f_d} = S_{z,H}^{x,f_{dH}} + S_{z,V}^{f_{dV}} \end{cases} \tag{15}$$

where $\varphi_f$ is interface friction angle; $q$ is additional frontal thrust on the working face; $\rho$ is the distance to the center of cutter head; $H$ is burial depth of tunnel center; $\mu$ is Poisson's ratio of soil.

The contour map of ground settlement and horizontal displacement caused by frictional force $f_d$ is shown in Figure 10a,b. The distributions of displacement show good regularity. Although, as the displacement values and action range are too small (less than 0.01 mm) even in deep soil quantitative analysis need not be carried out.

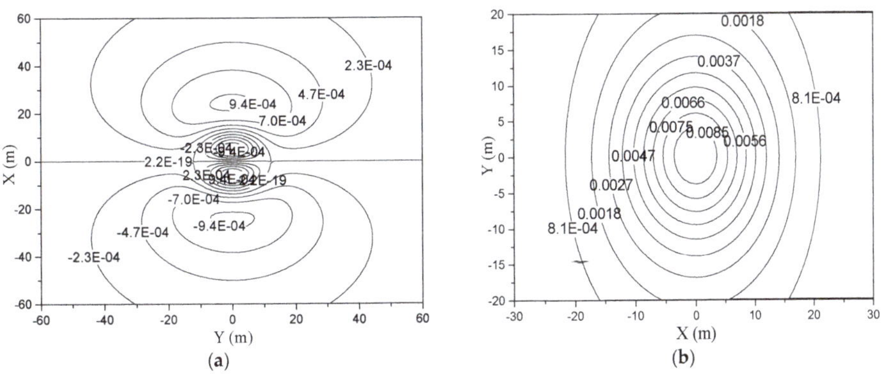


**Figure 10.** Ground movements caused by $f_d$: (**a**) Settlement; (**b**) Horizontal displacement.

### 4.4. Quantitative Analysis of Frictional Force $f_s$

As the characteristic of shell friction $f_s$ between the shield and the soil is like that between pile and soil, slip softening phenomenon should be considered in calculation of $f_s$ by Equation (15) [38–40]. Based on Equations (9) and (10) and the integral region in Figure 11, the displacement of $f_s$ can be calculated like $f_d$ above.

$$\begin{cases} f_s = \beta_s \sigma'_\theta \tan\delta \\ \sigma'_\theta = \sigma'_V \sin^2\theta + \sigma'_H \cos^2\theta \end{cases} \tag{16}$$

where $\beta_s$ is slip softening coefficient; $\sigma_\theta{}'$ is radial stress; $\delta$ is interface friction angle; $\sigma_V{}'$ is vertical stress; $\sigma_H{}'$ is horizontal stress.

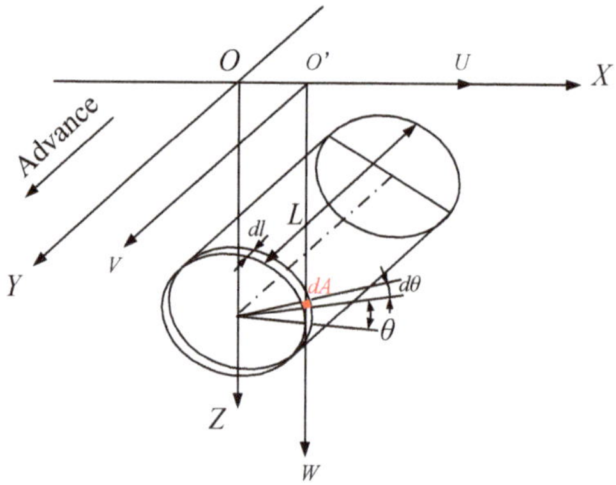


**Figure 11.** Integral region of $f_s$.

As we can see from Figure 12a,b, the ground settlement is symmetrically distributed along the Y axis, ground uplifts in front of the cutter head, settlements behind the cutter head. The horizontal displacement is symmetrically distributed along the diagonal line of X axis and Y axis.

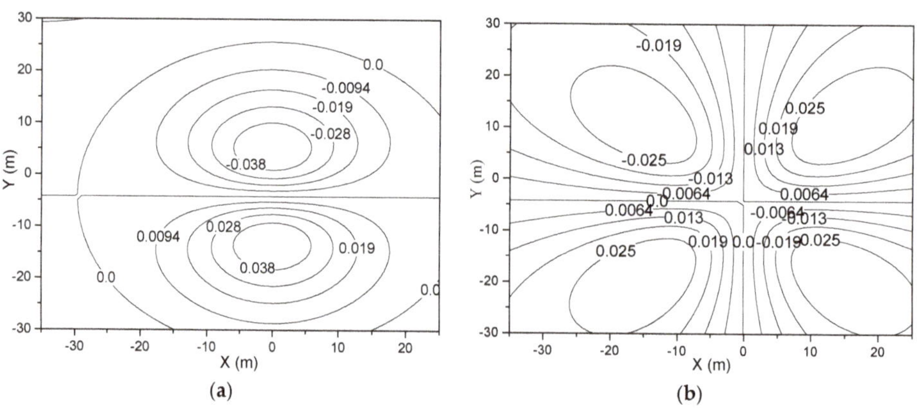


**Figure 12.** Ground movements caused by $f_s$: (**a**) Settlement; (**b**) Horizontal displacement.

### 4.5. Quantitative Analysis of Synchronous Grouting Pressure p

Synchronized grouting at the shield tail can fill the shield tail gap in time and reduce ground loss, but on the other hand, the grouting will extrude the surrounding soil. The extrusion action can be obvious with poor water permeability and high grouting pressure, which may cause stratum uplift.

In practice, the distribution pattern of synchronous grouting pressure is complex. In order to simplify the calculation, the following assumptions are made: the grouting pressure distributes uniformly in the circumference; little slurry diffuses into surrounding stratum in a short time, that is, the force on stratum is the same as grouting pressure; the longitudinal length of grouting is taken as 1 m [41]. Based on Equations (9) and (10) and the integral region in Figure 13, the displacement of $p$ can be calculated. As we can see from Figures 14 and 15, the displacement values and distribution range are small. It is worth noting that ground uplift increases significantly with the increase of grouting pressure in Figure 16, so grouting pressure needs to be selected reasonably.

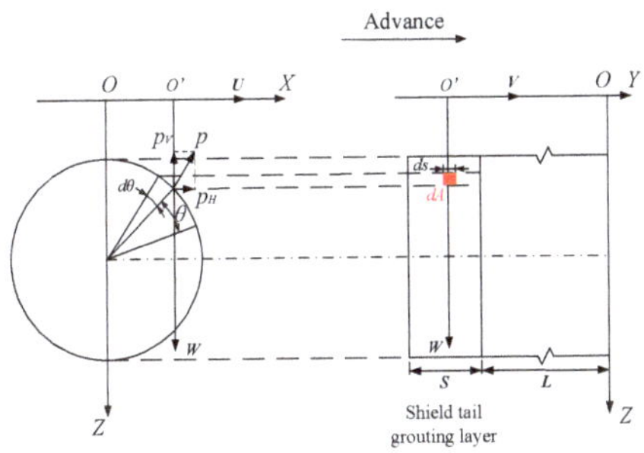


**Figure 13.** Integral region of $p$.

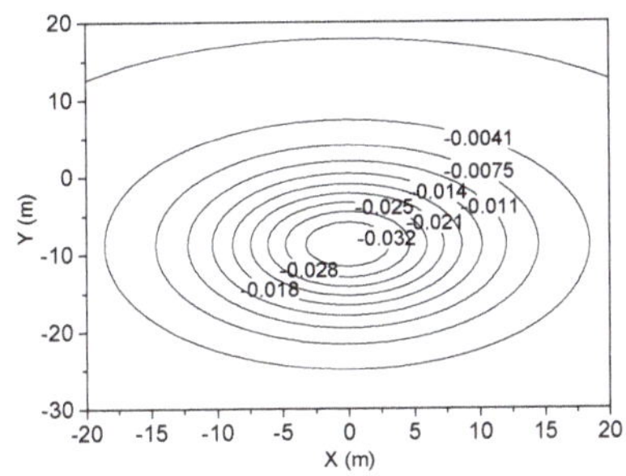


**Figure 14.** Ground uplift caused by $p$.

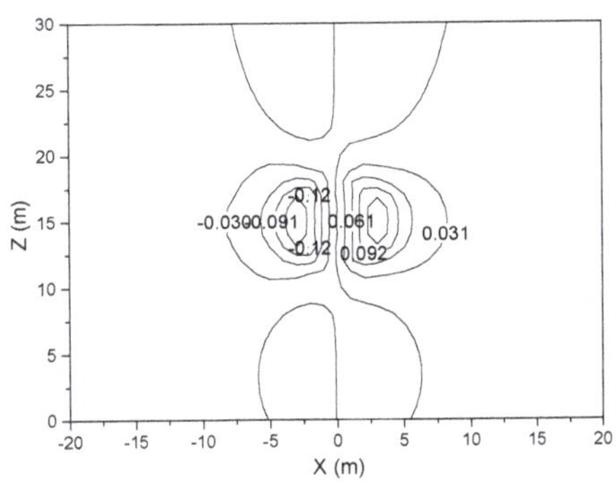


**Figure 15.** Horizontal displacement of stratum caused by $p$.

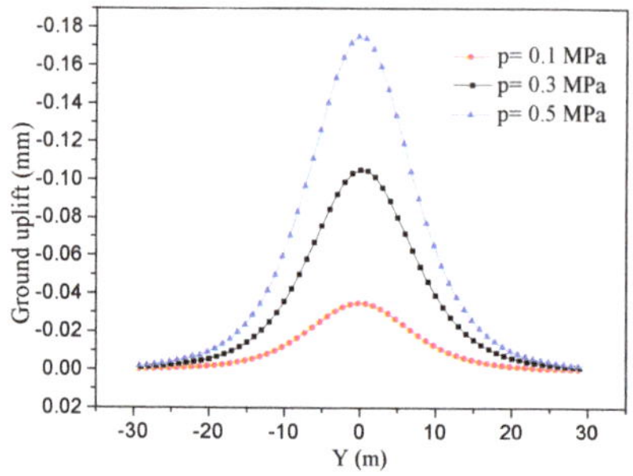


**Figure 16.** Ground uplift of different $p$.

## 5. Verification by Field Monitoring Data

The stratum displacements induced by various effect factors and total displacement are shown in Table 3. The effect ratio of ground loss is over 95%, which plays a decisive role. The ratios to total of frictional force $f_d$ and $f_s$ are much less than 1%; they are too small to be considered. The ratios of additional thrust $q$ and synchronous grouting pressure $p$ are a little larger, but are no more than 3%. As shown in Figures 8 and 16, the variations of these two factors make some small quantitative changes, but no qualitative differences. It shows that this conclusion has good universality for different working conditions. The total ground settlement is compared with field monitoring data in Figure 17. From this, we can see that there are some differences between them. For example, the calculating settlement maximum is 10.25 mm, while monitoring data is 9.7 mm. The difference rate is 5.3%, which belongs to acceptable error range. What is more, they have the same trend and the difference is stable. This is because that the elastic model is adopted in this paper, but in practice with the redistribution of stress, a pressure arch will be formed around the tunnel. The soil out of the arch is prevented from entering the loosening zone, which cause the field monitoring data to be slightly smaller than theoretical calculation result.

**Table 3.** The displacements induced by various effect factors.

| Effect Factors | Ground Settlement Maximum (mm) | Ratio to Total | Horizontal Displacement in Stratum (mm) | Ratio to Total |
|---|---|---|---|---|
| $q$ | −0.12 | 0.01 | 0.20 | 0.03 |
| $f_d$ | <0.01 | <0.01 | 0.04 | <0.01 |
| $f_s$ | ±0.04 | <0.01 | 0.03 | <0.01 |
| $p$ | −0.18 | 0.02 | 0.10 | 0.01 |
| Ground loss | 10.51 | 1.02 | 7.30 | 0.95 |
| total | 10.25 | 1.00 | 7.67 | 1.00 |

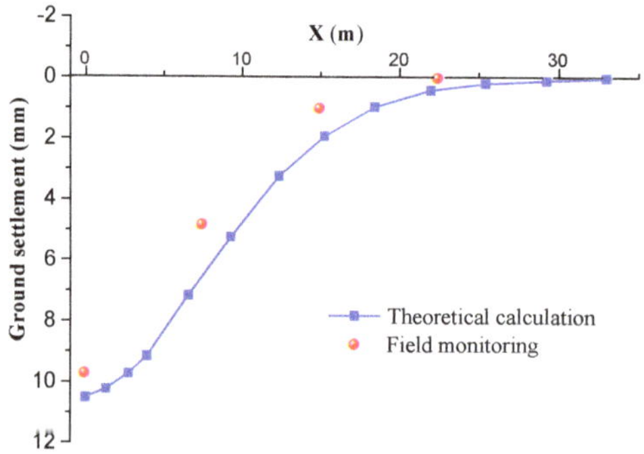


**Figure 17.** Total calculating displacement and field monitoring data.

## 6. Conclusions

In order to relieve the increasing ground traffic pressure in the process of urbanization in China, it is inevitable to build more metro lines. The evaluation and control of the stratum displacement is a prominent problem during shield construction of city metro engineers. Based on the engineering project of an interval tunnel between Shizishan Station and Chuanshi Station in Chengdu Metro Line 7, China, this paper analyzed the action mechanism of stratum displacement induced by shield tunneling, and divided effect factors into two categories: ground loss factors and mechanical factors. Through the quantitative analysis of the effect factors, the following conclusions can be drawn:

(1). The calculation results of this paper are applicable for shield tunneling in urban soft soil layer, but the calculation methods and thoughts have a good universality for different working conditions, it just needs necessary parameter adjustments. Combining the advantages of Loganathan method and mirror source-sink method, the three-dimensional displacement induced by ground loss are calculated. Based on the elastic half-space Mindlin model, the displacement at any point induced by four mechanical effect factors through coordinate transformation and different integral region is deduced. The total displacement is the sum of the two, which is verified by field monitoring data. The difference is stable and the difference rate is about 5.3%. This is because that the elastic model is adopted in this paper, but in practice with the redistribution of stress, a pressure arch will be formed around the tunnel. The soil out of the arch is prevented from entering the loosening zone, which causes the field monitoring data to be slightly smaller than theoretical calculation result. It is a methodological flaw, not a statistical one, and needs to be solved in future.

(2). The displacement induced by ground loss is over 95% of the total displacement, which plays a decisive role. The key to controlling stratum disturbance is decreasing the ground loss. When the outer diameter of shield and segment is determined, the most effective measure is decreasing the strain of grouting layer. As the stratum uplift caused by grouting pressure is much smaller than the settlement caused by grouting layer strain, the grouting pressure or grouting volume can be increased appropriately to ensure the grouting filling rate.

(3). The ratio of displacement caused by mechanical effect factors to total displacement is too small (less than 5%), even far less than the error caused by various simplified assumptions in theoretical calculation. In practice, this part of the displacement cannot be separated from the total displacement, it is difficult to verify its accuracy and validity by numerical simulation or field monitoring. Therefore, it is suggested that in calculation, the mechanical effect factors can be neglected as appropriate, and only the displacement caused by ground loss is considered, which can not only reduce the computational complexity and make it easy to use, but also can meet the requirements of engineering application accuracy.

**Author Contributions:** Y.Z. wrote the article; L.C. provided the parameters of project; H.Z. carried out the calculations and analyzed the calculation results; P.T. processed the field data; S.C. offered some useful suggestions for the preparation.

**Funding:** The study was supported by the National Natural Science Foundation of China (NSFC) under Grant No. 51508037.

**Acknowledgments:** We also highly appreciate the contribution of data collection from China Railway No. 9 Group Co., Ltd. Finally, the authors would like to thank reviewers for useful comments and editors for improving the manuscript.

**Conflicts of Interest:** The authors declare no conflict of interest.

## References

1. Zhang, H.; Chen, L.; Zhu, Y.; Zhou, Z.; Chen, S. Stress Field Distribution and Deformation Law of Large Deformation Tunnel Excavation in Soft Rock Mass. *Appl. Sci.* **2019**, *9*, 865. [CrossRef]
2. Zhu, Y.; Chen, L.; Zhang, H.; Zhou, Z.; Chen, S. Physical and Mechanical Characteristics of Soft Rock Tunnel and the Effect of Excavation on Supporting Structure. *Appl. Sci.* **2019**, *9*, 1517. [CrossRef]

3. Zeng, Y.H.; Liu, K.; Fang, Y.; Zhang, X.F.; Bai, Y. Fire Model Test on Temperature Field in the Rescue Station of an Extralong Railway Tunnel. *Adv. Civ. Eng.* **2019**, 1–12. [CrossRef]
4. Zhang, H.; Chen, L.; Chen, S.; Sun, J.; Yang, J. The spatiotemporal distribution law of microseismic events and rockburst characteristics of the deeply buried tunnel group. *Energies* **2018**, *11*, 3257. [CrossRef]
5. Lin, F.; Chen, S.G.; Zhang, H. Study on type selection of shield equipment in different geological conditions. In *Green Building, Materials and Civil Engineering*; CRC Press: Boca Raton, FL, USA, 2014; pp. 409–412.
6. Zhou, Z.L.; Chen, S.G.; Li, Y.S. Research on predicting and distribution of stratum displacement of double–tube parallel shield tunnel. *J. Highw. Transp. Res. Dev.* **2015**, *32*, 109–117.
7. Zou, J.F.; Qian, Z.H.; Xiang, X.H.; Chen, G.H. Face Stability of a tunnel excavated in saturated nonhomogeneous soils. *Tunn. Undergr. Space Technol.* **2019**, *83*, 1–17. [CrossRef]
8. Zhang, H.; Chen, S.G.; Deng, X.F. Analysis of the influence of shield driving parameters on ground settlements. *Mod. Tunn. Tech.* **2010**, *47*, 48–53.
9. Liao, S.M.; Liu, J.H.; Wang, R.L.; Li, Z.M. Shield tunneling and environment protection in Shanghai soft ground. *Tunn. Undergr. Space Technol.* **2009**, *24*, 454–465. [CrossRef]
10. Xie, X.Y.; Yang, Y.B.; Mei, J. Analysis of ground surface settlement induced by the construction of a large–diameter shield–driven tunnel in Shanghai, China. *Tunn. Undergr. Space Technol.* **2016**, *51*, 120–132. [CrossRef]
11. Zhang, H.; Chen, S.G.; Tan, X.R. Research on the mechanical behavior of a segment of a shield tunnel adjacent to a pile foundation. *Mod. Tunn. Tech.* **2012**, *49*, 101–107.
12. Zhang, H.; Chen, S.G.; Deng, X.F. Analysis on influence of shield tunneling on ground and bridge pile. *Chin. J. Underg. Sp. Eng.* **2011**, *7*, 552–557.
13. Peck, R.B. Deep Excavations and tunnelling in soft ground. In Proceedings of the 7th International Conference on Soil Mechanics and Foundation Engineering, Mexico City, Mexico, 25–29 August 1969; Balkema A A: Rotterdam, The Netherlands, 1969; pp. 225–290.
14. Celestino, T.B.; Gomes, R.A.M.; Bortolucci, A.A. Errors in group distortions due to settlement trough adjustment. *Tunn. Undergr. Space Technol.* **2000**, *15*, 97–100. [CrossRef]
15. Vorster, T.E.B.; Klar, A.; Soga, K.; Mair, R.J. Estimating the effects of tunneling on existing pipelines. *J. Geotech. Geoenviron.* **2005**, *131*, 1399–1410. [CrossRef]
16. Attewell, P.B.; Yeates, J.; Selby, A.R. *Soil Movements Induced by Tunnelling and Their Effects on Pipelines and Structures*; Thomson Science and Professional: 2nd Floor Aldgate House, 33 Aldgate High Street, London, UK, 1986.
17. Osman, A.S.; Bolton, M.D.; Mair, R.J. Predicting 2D ground movement around tunnels in undrained clay. *Geotech* **2006**, *56*, 597–604. [CrossRef]
18. New, B.M.; O'Reilly, M.P. Tunneling–induced ground movements, predicting their magnitude and effects. In Proceedings of the 4th Conference on Ground Movements and Structures; Pentech Press: London, UK, 1992; pp. 671–697.
19. O' Reilly, M.P.; New, B.M. Settlement above tunnels in the United Kingdom their magnitude and prediction. In Proceedings of the Tunneling' 82 Symposium, London, UK, 7–11 June 1982; pp. 173–181.
20. Sagaseta, C. Analysis of undrained soil deformation due to ground loss. *Geotechnique* **1987**, *37*, 301–320. [CrossRef]
21. Loganathan, N.; Poulos, H.G. Analytical prediction for tunneling–induced ground movements in clays. *J. Geotech. Geoenviron.* **1998**, *124*, 846–856. [CrossRef]
22. Federico, P.; Andrew, J.W. Ground movements due to shallow tunnels in soft ground: 1. analytical solutions. *J. Geotech. Geoenviron.* **2013**, *140*, 04013040.
23. Kung, G.T.; Juang, C.H.; Hsiao, E.C.; Hashash, Y.M. Simplified model for wall deflection and ground–surface settlement caused by braced excavation in clays. *J. Geotech. Geoenviron.* **2007**, *133*, 731–747. [CrossRef]
24. Peng, F.L.; Wang, H.L.; Tan, Y.X.; Zheng, L.; Li, Y.L. Field measurements and finite–element method simulation of a tunnel shaft constructed by pneumatic caisson method in shanghai soft ground. *J. Geotech. Geoenviron.* **2011**, *137*, 516–524. [CrossRef]
25. Chakeri, H.; Ozcelik, Y.; Unver, B. Effects of important factors on surface settlement prediction for metro tunnel excavated by EPB. *Tunn. Undergr. Space Technol.* **2013**, *36*, 14–23. [CrossRef]
26. Zhang, Z.G.; Huang, M.S. Geotechnical influence on existing subway tunnels induced by multiline tunneling in Shanghai soft soil. *Comput. Geotech.* **2014**, *56*, 121–132. [CrossRef]

27. Son, M.; Cording, E.J. Estimation of building damage due to excavation–induced ground movements. *J. Geotech. Geoenviron.* **2005**, *131*, 162–177. [CrossRef]
28. Ahmed, M.; Iskander, M. Analysis of tunneling–induced ground movements using transparent soil models. *J. Geotech. Geoenviron.* **2010**, *137*, 525–535. [CrossRef]
29. Mindlin, R.D. Force at a point in the interior of a semiinfinite solid. *Physics* **1936**, *7*, 195–202. [CrossRef]
30. Huang, M.S.; Zhang, C.R.; Li, Z. A simplified analysis method for the influence of tunneling on grouped piles. *Tunn. Undergr. Space Technol.* **2009**, *24*, 410–422. [CrossRef]
31. Zhang, Z.G.; Huang, M.S.; Zhang, M.X. Theoretical prediction of ground movements induced by tunneling in multilayered soils. *Tunn. Undergr. Space Technol.* **2011**, *26*, 345–355. [CrossRef]
32. Sun, H.S.; Lei, G.H.; Ng, C.W.W. Displacements under linearly distributed pressures by extended Mindlin's equations. *Comput. Geotech.* **2013**, *50*, 143–149. [CrossRef]
33. Antonio, B. Analytical solutions for shallow tunnels in saturated ground. *J. Eng. Mech.* **2001**, *127*, 1258–1266.
34. Verruijt, A.; Booker, J.R. Surface settlements due to deformation of a tunnel in an elastic half plane. *Geotechnique* **1996**, *46*, 753–756. [CrossRef]
35. Verruijt, A. Deformations of an elastic half plane with a circular cavity. *Int. J. Solids Struct.* **1998**, *35*, 2795–2804. [CrossRef]
36. Lee, K.M.; Rowe, R.K.; Lo, K.Y. Subsidence owing to tunneling I: Estimating the gap parameter. *Can. Geotech. J.* **1992**, *29*, 929–940. [CrossRef]
37. Zhang, Q.Q.; Li, S.C.; Li, L.P.; Chen, Y.J. Simplified method for settlement prediction of pile groups considering skin friction softing and end resistance hardening. *Chin. J. Rock Mech. Eng.* **2013**, *32*, 615–624.
38. Potyondy, J.G. Skin friction between various soils and construction materials. *Geotechnique* **1961**, *11*, 339–353. [CrossRef]
39. Alonso, E.E.; Josa, A.; Iedesma, A. Negative skin friction on Piles: A simplified analysis and prediction procedure. *Geotechnique* **1984**, *34*, 341–357. [CrossRef]
40. Wang, H.X. Effect of cutterhead compressing the front soil and influence of head aperture ratio on contact pressure of EPB shield to the front soil. *Chin. Civ. Eng. J.* **2009**, *42*, 113–118.
41. Li, Z.M.; Liao, S.M.; Dai, Z.R. Theoretical study on synchronous grouting filling patterns and pressure distribution of EPB shield tunnels. *Chin. J. Geotech. Eng.* **2010**, *32*, 1752–1757.

applied sciences

MDPI

Article

# An Improved Method for Predicting the Greenfield Stratum Movements Caused by Shield Tunnel Construction

XinRong Tan [1], Heng Zhang [2,*], Gang Zhang [2], Yimo Zhu [2,*] and Peng Tu [3]

1 China Railway Eryuan Engineering Croup Co. LTD, Chengdu 610031, China; xy.tan@163.com
2 Key Laboratory of Transportation Tunnel Engineering, Ministry of Education, Southwest Jiaotong University, Chengdu 610031, China; zg19970521@my.swjtu.edu.cn
3 Department of Road and Bridge Engineering, Sichuan Vocational and Technical College of Communications, Chengdu 611130, China; pengtu1990@outlook.com
* Correspondence: Hengzhang1985@outlook.com (H.Z.); zhuyimo1994@163.com (Y.Z.); Tel.: +86-028-8763-4386 (H.Z. & Y.Z.)

Received: 23 September 2019; Accepted: 21 October 2019; Published: 24 October 2019

**Abstract:** Shield tunneling is becoming the preferred construction scheme for metro construction because of its advantages of fast construction speed and small disturbance. However, limited by process defects, the stratum movements induced by the construction of shield tunnels still affects the safety of nearby underground structures and aboveground buildings. Therefore, the reliable prediction of stratum movements is important. Described in this paper is an analysis method of the Greenfield stratum movements (Greenfield is an area of land that has not yet had buildings on it, stratum movements means the movement of various soil layers) caused by shield tunnel construction combining an elastic half-space model of mirror source–sink method with the use of modified analytical method. Based on the theoretical formula in this paper, not only can the curve of surface settlement trough be calculated, but also the three-dimensional displacement field of deep soil can be obtained. By comparing vertical and horizontal contour maps of Greenfield stratum movements, good consistency between theoretical formula results and centrifugal test results are shown. This solves the defects and limitations of existing two-dimensional formulas; furthermore, based on this, it is convenient to evaluate the effect on the other skewed underground structures through the elastic foundation beam and other similar methods; therefore, this paper can provide a wide guidance and service for the design and construction of underground engineering in the future.

**Keywords:** shield tunnel; stratum movements; analytical; mirror source–sink method; centrifuge modelling test

## 1. Introduction

Since the 21st century, tunnel and underground engineering in China have made great progress [1–7]. With the continuous development of urbanization in China, the urban population is growing, and urban traffic pressure is also increasing. In order to alleviate the urban traffic pressure, city managers have taken a lot of measures, such as limited number driving, expansion, and transformation of existing roads, but the traffic demand is still far greater than the current urban ground carrying capacity. Metro has the advantages of large passenger capacity, fast speed, and full use of underground space, which is more and more popular in modern society. However, the urban subway has strict requirements for its construction methods. Because of the high density of buildings in cities, the construction of the metro will cause stratum movement and bring adverse effects on ground buildings. Especially when passing through high-rise buildings, the disturbance caused by construction on stratum must be more strictly controlled. The construction period of Metro is long, and the interruption of traffic should be avoided as far as possible when crossing the road. Usually, subway construction methods include the open-cut method, shallow-buried excavation method, and shield tunneling method. Among them, the shield tunneling method has become the preferred construction scheme for metro construction because of its great advantages of fast construction speed, a high degree of automation, and small disturbance to stratum [8,9].

Although shield tunneling has many advantages mentioned above, and the construction technology has made great progress after many years of development, due to the defects of geological conditions and construction technology, the advance of shield tunneling will inevitably make a disturbance, change the stress state of soil, and cause stratum displacement, as shown in Figure 1. If the disturbance of stratum cannot be clearly understood, it will affect the safety of nearby underground structures and aboveground buildings. For example, the adjacent metro tunnels may cause deformation restrictions beyond the normal operation of the metro, or even lead to train derailment; it may also lead to the breakage of underground pipelines, resulting in a series of problems, such as gas leakage, interruption of urban water use, interruption of communication power system, and so on, affecting the daily life of the city [10–12]. Therefore, a reliable prediction of stratum movements is important.

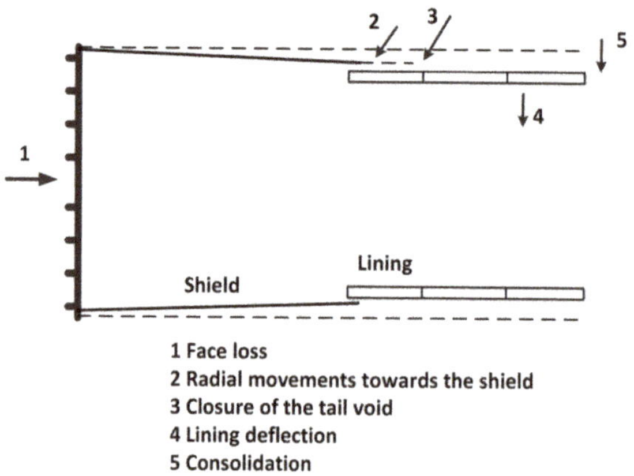


**Figure 1.** Sources of stratum movements of shield tunnel.

Over the years, previous researchers have done numerous studies on the effects of shield tunnel construction on stratum displacement [13–18]. In the direction of theoretical research, Peck assumed that the curve of land settlement trough satisfies the conditions of Gauss distribution and invariant stratum volume, and deduced Peck's empirical formula based on many field monitoring data [19]. Attewell et al. used Peck's empirical formula to simulate the free displacement of soil at the location of existing tunnels [20–24], and then deduced the displacement of existing tunnels with elastic foundation beams and other models. However, Celestino and Klar [25,26] gradually found that the Peck curve could not accurately describe soil settlement trough in many cases. Voster, O'Reilly et al. [27–30] have proposed fitting curves from different research angles to fit the free displacement field of soils. Litwinszyn [31] put forward the theory of random medium through a sand box test. Liu, Yang, et al. [32–34] introduced this theory into the prediction of stratum displacement caused by the excavation of geotechnical tunnels. Although these studies have the advantages of concise calculation formula and convenient use, they lack a clear theoretical basis of mechanics. They can only be called a mathematical empirical method, not an analytical solution of soil settlement. Because the empirical formula is fitted by a large number of field monitoring data, the parameters of the fitting function are closely related to the actual construction conditions, and lack of a clear theoretical basis, the accuracy of the results predicted by this method is often difficult to meet the requirements, and its defects are evident. Comparatively, the solving process of the analytic method and modified analytic method is relatively complex, but through more rigorous analysis, as long as the parameters are selected properly, it has better universality. Sagaseta [35] deduced the expression of three-dimensional displacement of stratum by using the mirror source–sink method. This formula can not only calculate the curve of surface settlement trough, but also calculate the displacement and stress of deep soil, so it is widely used [36–43]. However, the movement pattern of actual volume loss is not consistent with the assumed equivalent radial movement pattern in this method and has some deviation. Based on field monitoring data, Loganathan, Jiang, and other researchers [44–51] modified the stratum displacement pattern by exponential function and deduced the prediction formula of surface subsidence under non-equivalent radial movement pattern. However, most of these methods are only suitable for three-dimensional orthogonal plane or two-dimensional cases, which have great defects and need to be improved urgently. Many other researchers concerned the field description, soil classification for engineering purposes [52–54], and the analysis methods of data such as neural networks [55–58]. Saplachidi et al. [59] used measured settlements from the tunneling excavation for the extension of Line 3 of the Athens Metro for the verification and calibration of the empirical formulae. Suwansawat et al. [60] evaluated the potential and the limitations of artificial neural networks (ANN) for predicting surface settlements caused by Earth pressure balance (EPB) shield tunneling and to develop optimal neural network models for this objective. Ahangari et al. [61] created a database from previous research [62–65] and studied the capability of adaptive neuro-fuzzy inference system (ANFIS) and gene expression programming (GEP) methods for settlement prediction. However, those methods belong to the fuzzy solutions without mechanism analysis, training neural networks requires a lot of data from databases or numerical simulation, and there are problems such as difficult convergence or slow convergence speed.

This paper studies the Greenfield stratum movements caused by shield tunnel construction combining an elastic half-space model of mirror source–sink method with the use of modified analytical method. Through theoretical derivation and empirical function modification, the prediction formula of three-dimensional Greenfield stratum movements is improved, and it shows good consistency with the centrifugal test results. The new solution can calculate the three-dimensional stratum displacement accurately, which is unavailable for the previous methods. It can be used in the process of tunnel excavation rather than only for the ultimate settlement. Based on this improved solution, it is convenient to evaluate the effect on the other skewed underground structures through the elastic foundation beam and other similar methods; therefore, this paper can provide a wide guidance and service for the design and construction of underground engineering in the future.

## 2. Basic Theory and Calculation Method

### 2.1. Disturbance Factors

Usually, there are three factors that cause disturbance of surrounding soil by shield tunneling technology: First, the gap between segment lining and stratum due to the reasons of assembling segment lining, shield overcutting, shield snake-shaped, and shield body adjustment; although synchronous grouting is adopted, the grouting cannot completely eliminate the gap at present [66]; the second is the elastic-plastic deformation of the soil at the shield heading face, which is closely related to the soil characteristics and the additional force on the shield working face. Thirdly is the horizontal friction between the outer surface of the shield and the surrounding soil. Compared with the effect of the gap, the stratum movements caused by the additional force on the shield working face and the frictional force between the outer surface of the shield and the surrounding soil during tunnel construction is very small [67]. Therefore, this paper neglects the effects of the latter two factors when analyzing, and mainly studies the effects of the gap. Sources of stratum movements of shield tunnel are shown in Figure 1 and the coordinate system is set as shown in Figure 2 [68].

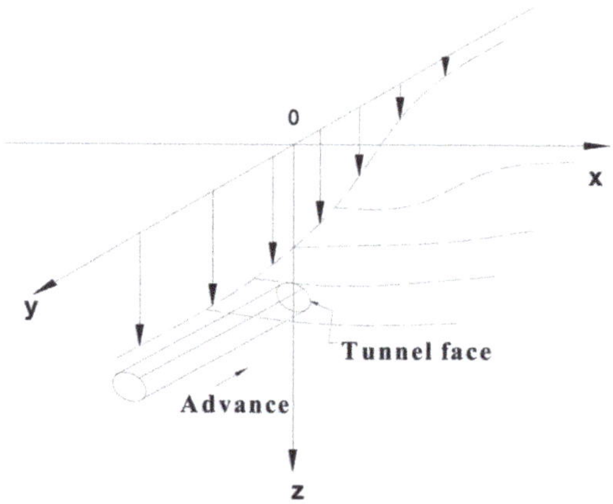


**Figure 2.** Surface displacement caused by shield tunnel.

### 2.2. Gap Parameter

In order to analyze the effects of the above factors, Rowe et al. [69,70] proposed the concept of gap parameter to reflect the stratum loss caused by shield tunnel construction, transforming the three-dimensional problem to a two-dimensional gap in plane. The composition of the gap takes into account the following factors: Elastic-plastic deformation of the shield at the tunnel excavation face, the over-excavation of the shield machine, the physical gap between the outer skin of shield shell and segment lining, and the construction technology. The formula for calculating the gap parameters $g$ under undrained condition is as in Equation (1).

$$g = G_P + U_{3D}^{*} + \omega \quad (1)$$

where shows that the gap parameters are composed of three parts [44]:

The first part $G_p$ is the physical gap, which represents the gap between the outer skin of shield shell and segment lining; its expression is as in Equation (2):

$$G_P = 2\Delta + \delta \quad (2)$$

where $\Delta$ is the thickness of shield shell and $\delta$ is the space formed by segment lining installation.

The second part $U_{3D}$ is the elastic-plastic deformation of shield at the tunnel excavation face; its quantitative expressions are as in Equations (3) and (4):

$$U_{3D} = \frac{k}{2}\delta_y = \frac{k}{2}\frac{\Omega R P_0}{E} \quad (3)$$

$$P_0 = K_0' P_v' + P_w - P_f \quad (4)$$

where $k$ is the soil cutting resistance coefficient; $\Omega$ is the displacement coefficient; $R$ is the tunnel radius; $E$ is the elastic modulus; $K'_0$ is the effective static lateral pressure coefficient; $P'_v$ is the effective vertical stress at depth of tunnel center; $P_w$ is the pore water pressure at depth of tunnel center; and $P_f$ is the supporting force provided by tunnel.

The third part $\omega$ is the shield construction technology coefficient, of which the value is as in Equation (5):

$$\begin{cases} \frac{U_i}{R} = 1 - \left( \frac{1}{1 + \frac{2(1+\mu_u)C_u}{E_u} \left[ exp\left(\frac{N-1}{2}\right) \right]^2} \right)^{\frac{1}{2}} \\ N = \frac{\gamma H - p_f}{C_u} \\ \omega = min\left\{ 0.6G_p, \frac{1}{3}U_i \right\} \end{cases} \quad (5)$$

where $U_i$ is the elastoplastic plane strain displacement at the tunnel crown; $\mu_u$ is the soil Poisson's ratio under undrained condition; $C_u$ is the soil shear strength under undrained condition; $E_u$ is the soil elastic modulus under undrained condition; $\gamma$ is the unit weight of soil; and $H$ is the buried depth of tunnel center.

## 2.3. Loganathan' Solution

Based on the above formulas, Loganathan proposed to use plane loss rate $\varepsilon_{x,z}$ to quantify stratum loss with non-equivalent radial movement. The expression is as in Equation (6):

$$\varepsilon_{x,z} = \frac{4gR + g^2}{4R^2} \exp\left\{ - \left| \frac{1.38x^2}{(H+R)^2} + \frac{0.69z^2}{H^2} \right| \right\}. \quad (6)$$

Combined with Verruijit's [71,72] analytical formula, he put forward a formula for estimating stratum displacement caused by undrained stratum loss in clays, of which the formulas are as in Equation (7):

$$\begin{cases} S_z = R^2 \left\{ -\frac{z-H}{x^2+(z-H)^2} + (3-4\mu)\frac{z+H}{x^2+(z+H)^2} - \frac{2z\left[x^2-(z+H)^2\right]}{\left[x^2+(z+H)^2\right]^2} \right\} \times \varepsilon_{x,z} \\ S_x = -R^2 x \left\{ \frac{1}{x^2+(z-H)^2} + \frac{(3-4\mu_u)}{x^2+(z+H)^2} - \frac{4z(z+H)}{\left[x^2+(z+H)^2\right]^2} \right\} \times \varepsilon_{x,z} \end{cases} . \quad (7)$$

Considering that the displacement solution caused by stratum loss derived by Loganathan is a two-dimensional solution, which is often a three-dimensional problem in actual engineering, it is necessary to extend the two-dimensional displacement solution to three-dimensional. In the above deduction process, we can find that the stratum loss parameter $\varepsilon_{x,z}$ is the key for extending. If we extend the plane stratum loss parameter $\varepsilon_{x,z}$ to three-dimensional parameter $\varepsilon_{x,z,y}$, the displacement solution can be extended to three-dimensional space. Another important method to solve the stratum displacement caused by stratum loss is used here. The concept of stratum loss proposed by Sagaseta proposed the mirror source–sink method and derives the problem of free displacement field by elastic half-space model. Although field monitoring data [40] show that the movement pattern of actual volume loss is not consistent with the assumed equivalent radial movement pattern in this method, its process is a strict mathematical derivation process, so the three-dimensional stratum loss parameter $\varepsilon_{x,z,y}$ can be extracted from the Sagaseta solution.

## 2.4. Sagaseta's Solution

Sagaseta used the mirror source–sink method, assumed that the surface is free, the soil is isotropic, and incompressible. The analysis steps are as follows and in Figure 3:

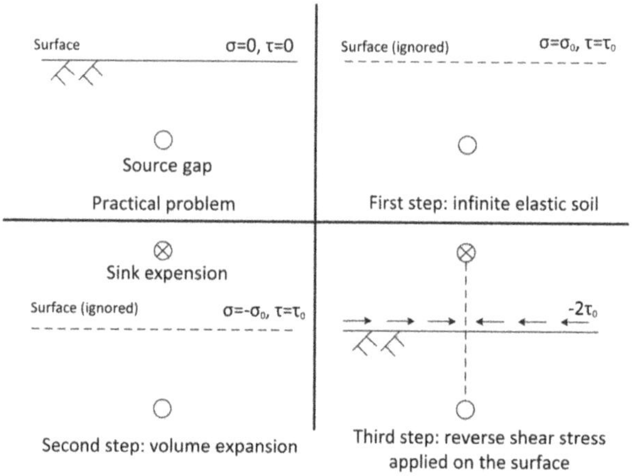


**Figure 3.** Analysis steps of mirror source–sink method.

First step: The existence of the original surface is ignored, and the half-space problem is transformed into the internal gap problem of full space. Under the action of a source gap, stress field changing, there will be the normal stress $\sigma_0$ and the shear stress $\tau_0$ at the original surface position.

Second step: A sink is set at the mirror position of the source gap above the original surface; there is a volume expansion of the same size at the sink position. The sink expansion will produce a normal stress $-\sigma_0$ of the same value and opposite direction and a shear stress $\tau_0$ of the same value and same direction at the original surface position. Therefore, under the action of source and sink, the normal stress on the original surface can be offset, which does not exist in practice.

Third step: Under the action of source and sink, a shear stress $-2\tau_0$ of twice the value and opposite direction is supposed at the original surface position and the corresponding displacement field of the surface shear stress below the surface at each point can be obtained.

Final step: The displacements generated by the above three steps at any point below the surface are $S_1$, $S_2$, and $S_3$, respectively. The displacement solution of the practical problem is $S = S_1 + S_2 + S_3$.

Based on above steps, Sagaseta deduced the expression of surface displacement caused by tunnel construction is as in Equation (8):

$$\begin{cases} S_{x0} = -\dfrac{v_{loss}}{2\pi}\dfrac{x}{x^2+H^2}\left[1+\dfrac{y}{(x^2+y^2+H^2)^{\frac{1}{2}}}\right] \\ S_{y0} = \dfrac{v_{loss}}{2\pi}\dfrac{1}{(x^2+y^2+H^2)^{\frac{1}{2}}} \\ S_{z0} = \dfrac{v_{loss}}{2\pi}\dfrac{H}{x^2+H^2}\left[1+\dfrac{y}{(x^2+y^2+H^2)^{\frac{1}{2}}}\right] \\ S_{x0}(y \to \infty) = -\dfrac{v_{loss}}{\pi}\dfrac{x}{x^2+H^2} \\ S_{z0}(y \to \infty) = \dfrac{v_{loss}}{\pi}\dfrac{H}{x^2+H^2} \end{cases} \quad (8)$$

where $V_{loss}$ is stratum loss. Contrastive Equation (8) can be found:

$$\begin{cases} S_{x0}(y) = S_{x0}(y \to \infty) \times \frac{1}{2} \times \left[ 1 + \frac{y}{(x^2+y^2+H^2)^{\frac{1}{2}}} \right] \\ S_{z0}(y) = S_{z0}(y \to \infty) \times \frac{1}{2} \times \left[ 1 + \frac{y}{(x^2+y^2+H^2)^{\frac{1}{2}}} \right] \end{cases} . \quad (9)$$

## 2.5. An Improved Method

Equation (9) shows that the distribution of stratum displacement caused by shield construction (stratum loss) along the y-axis driving satisfies the law of $\frac{1}{2} \times \left[ 1 + \frac{y}{(x^2+y^2+H^2)^{\frac{1}{2}}} \right]$. Therefore, based on this law, the Loganathan's solution is extended to the three-dimensional. It is noticed that the coordinate system adopted in this paper is different from that used in the Sagaseta's solution, so the distribution law should be revised, and its expression is as in Equation (10):

$$\varepsilon_y = \frac{1}{2} \times \left[ 1 - \frac{y}{(x^2+y^2+H^2)^{\frac{1}{2}}} \right]. \quad (10)$$

The plane stratum loss parameter $\varepsilon_{x,z}$ is extended to three-dimensional parameter $\varepsilon_{x,z,y}$:

$$\varepsilon_{x,y,x} = \varepsilon_y \times \varepsilon_{x,z} = \left[ 1 - \frac{y}{(x^2+y^2+H^2)^{\frac{1}{2}}} \right] \times \frac{4gR+g^2}{8R^2} \exp\left\{ - \left[ \frac{1.38x^2}{(H+R)^2} + \frac{0.69z^2}{H^2} \right] \right\} \quad (11)$$

$\varepsilon_{x,z}$ is replaced by $\varepsilon_{x,z,y}$ in Equation (7) and Equation (12) is the three-dimensional analytical solution of stratum displacement caused by stratum loss. When $y = 0$, $z = 0$, the curves of settlement trough calculated by Equations (7), (8), and (12) are the same, and Equation (8) is the particular solution about the surface settlement value of Equation (12) at the tunnel face.

$$\begin{cases} S_z = R^2 \left\{ -\frac{z-H}{x^2+(z-H)^2} + (3-4\mu)\frac{z+H}{x^2+(z+H)^2} - \frac{2z\left[x^2-(z+H)^2\right]}{\left[x^2+(z+H)^2\right]^2} \right\} \times \varepsilon_{x,y,z} \\ S_x = -R^2 x \left\{ \frac{1}{x^2+(z-H)^2} + \frac{(3-4\mu)}{x^2+(z+H)^2} - \frac{4z(z+H)}{\left[x^2+(z+H)^2\right]^2} \right\} \times \varepsilon_{x,y,z} \end{cases} \quad (12)$$

For example, let $H = 13.65$ m, $\mu = 0.4$, $R = 2.325$ m, $g = 0.058$ m, then the displacement of soil layers at $z = 0$ m and $z = 10$ m is showed in Figure 4.

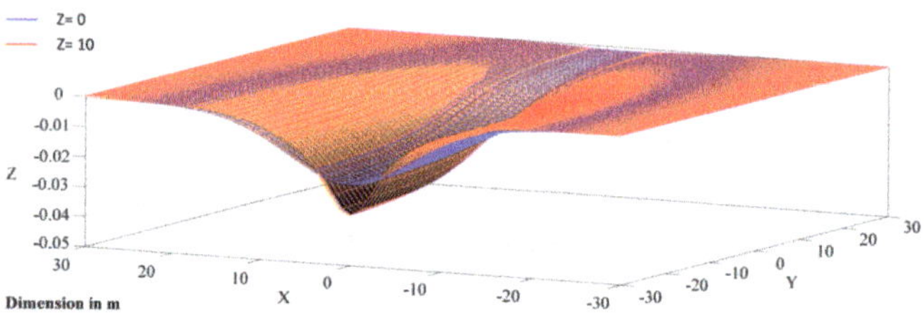


**Figure 4.** Settlement trough in different depth stratum (example).

## 3. Compared with Centrifugal Test Results

### 3.1. Material Properties

British scholar Marshall [73] designed a series of model tests to study the influence of shield tunneling in sandy ground by using the geocentrifugal testing machine of Cambridge University. The design conditions and dimensions of centrifuge model test are shown in Figure 5. The bottom dimension of the test box containing model tunnels and sand is 770 mm × 147.5 mm, and the sand depth after filling is 311 mm; the axis depth of the shield tunnel model is $Z_t$ = 182 mm, the tunnel diameter is $D_t$ = 62 mm, the section dimension of the model tunnel is shown in Figure 6, and the length of the shield tunneling section is $L_0$ = 147.5 mm.

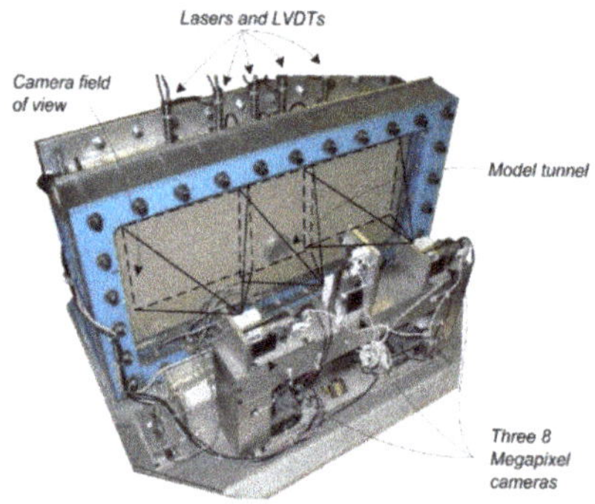


**Figure 5.** Centrifuge package—tunneling in sandy ground.

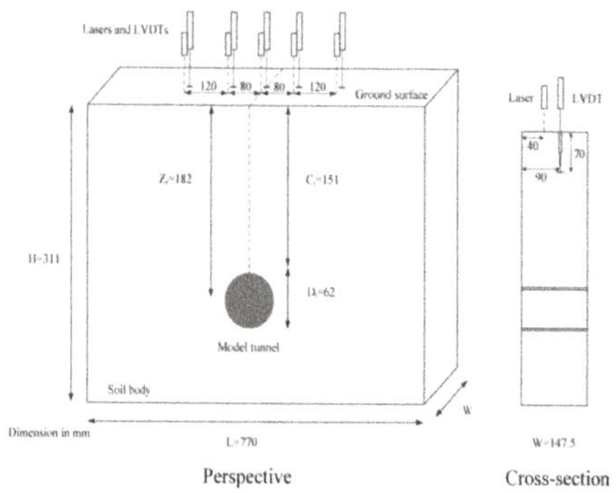


**Figure 6.** Centrifuge package schematic where LVDTs (linear variable differential transformers) were used to measure vertical sub-surface soil displacements at the same locations as the lasers. Cameras were used to capture images of the soil body during tests for analysis.

The test soil samples were taken from dry Leighton Buzzard Fraction E silica sand, UK, and its parameters are shown in Table 1. This sand has a typical $D_{50}$ of 122 μm [74], a specific gravity of 2.67, and maximum and minimum void ratios of 0.97 and 0.64 [75], respectively. Uniform sand with relative density of 90% was prepared by automatic sand leaker. The elastic modulus $E$ of sand soil is 20 MPa and Poisson's ratio $\mu$ is 0.4. The whole test process was carried out under the centrifugal acceleration of 75× $g$ (g is the acceleration of gravity); namely, the scale ratio of the test size was 75 times. Therefore, this test is equivalent to simulating a shield tunnel with a diameter of 4.65 m (model diameter 62 mm) and a tunnel central burial depth of 13.65 m (model tunnel central burial depth 182 mm) passing through Greenfield in reality. In the process of shield excavation, the simulation of stratum loss is realized by using a cylinder sealed with liquid connected to the tunnel driven by a motor. The stratum loss rate rises at a rate of about 0.3% per minute. The above test parameters are brought into Equation (12) for theoretical calculation of stratum displacement. Considering the clarity and regularity of result image, the test results of medium volume loss of 2.5% are compared.

**Table 1.** Physical parameters of soil samples.

| Material | Unit Weight (kN/m$^3$) | Typical $D_{50}$ (μm) | Maximum/Minimum Void Ratios | Relative Density | Poisson's Ratio | Elastic Modules (MPa) |
|---|---|---|---|---|---|---|
| Dry Leighton Buzzard Fraction E silica sand, UK. | 26.7 | 122 | 0.97/0.64 | 90% | 0.4 | 20 |

## 3.2. Comparative Analysis of Results

The comparative analysis of the results is shown in Figure 7; the vertical displacement contours of 37.5 mm and 30 mm in the central part show good consistency and regularity. In general, the centrifugal test results are nearly straight-line contours, while the theoretical calculation results are arc contours, which also leads to a large difference between 22.5 mm, 15 mm, and 7.5 mm contours, and shows a trend of increasing the difference from the center to the edge.

As for vertical displacement, the numerical value is relatively small, and its regularity is not obvious. This can also be seen from the poor symmetry of the left and right halves of the horizontal displacement in the centrifugal test nephogram. The shape and distribution position of centrifugal contours and theoretical contours are similar, especially the upper 4 mm curve and the central 2 mm curve. However, the difference between theoretical contours and centrifugal contours also shows an increasing trend from the center to the edge, the whole lines shift to the right a little, and the theoretical contour has a wider range. This is consistent with the law that the settlement trough of sand is narrower and deeper than that of clay [76].

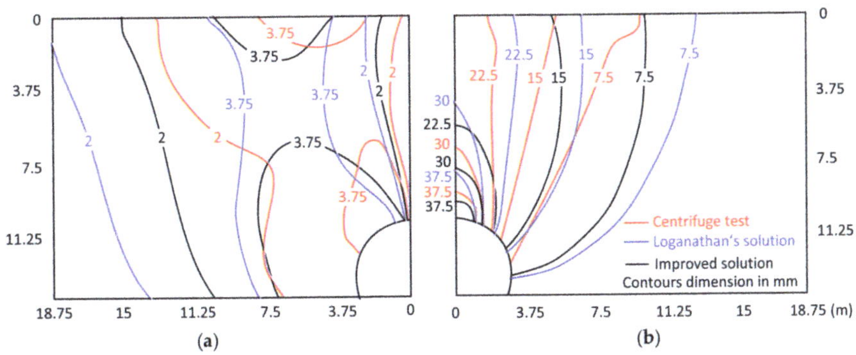


**Figure 7.** Contour comparison of Loganathan's solution, theorical calculation of this paper, and centrifugal test results (2.5% volume loss): (**a**) Horizontal displacement contours for Greenfield; (**b**) vertical displacement contours for Greenfield.

Due to sand soil being chosen as the test soil sample in centrifugal test, Sagaseta's solution is applicable to sand soil, but another important part of our theoretical formula, Loganathan's solution, is only applicable to clay soil. Be aware of the exponential function of Loganathan' solution modify the formation loss coefficient by fitting the clay soil data, although the soil parameters substituted in Equation (12) are sand soil's, which cannot eliminate the difference between the Loganathan's $\varepsilon_{x,z}$ and the properties of sand. This leads to the fact that in the vertical displacement contours, the centrifugal contours are linear and the theoretical contours arc, which is similar to the difference between the two on the slip fracture surface of soil. Generally, clay soil has a complex composition, elastic-plastic deformation occurs easily, deformation has a time-delay property, and many other disturbing factors, all of which are very unfavorable to the test analysis. However, sand soil has a single material, of which properties are easier to control in the test, and the drainage or non-drainage conditions need not be considered. Therefore, the current centrifugal tests are mostly based on sand. We regret that we have not found a suitable clay centrifugal test as a reference. However, considering the inevitable errors in centrifugal test and theoretical calculations, the comparison results are still acceptable.

## 4. Conclusions

(1) A new method for predicting the stratum movements caused by EPB shield tunnel construction is developed that combines an elastic half-space model of mirror source–sink method with the use of modified analytical method. Compared with the effect of a physical gap, the stratum movements caused by the additional force on the shield working face and the frictional force between the outer surface of the shield and the surrounding soil during tunnel construction is very small. Therefore, this paper neglects the effects of the latter two factors when analyzing, and mainly studies the effects of the physical gap. The method also takes into account non-equivalent radial ground loss.

(2) Sagaseta's method assuming equivalent radial movement pattern can only obtain the surface movements from the tunnel excavation. Considering non-equivalent radial displacement model, Loganathan's solution modifies Verruijit's analytical formula by exponential function, but it is only applicable to plane strain cases. The two methods have great limitations. For example, they cannot show the soil displacement of the x = y plane. If there is a new shield tunnel skewed with an old tunnel or pipeline and the displacement effect at the old tunnel or pipeline needs to be calculated, neither of the above two methods will apply. This paper combines the advantages of the two methods to derive a formula for predicting the three-dimensional stratum displacement caused by shield tunneling.

(3) Vertical displacement contours in the central part show good consistency and regularity, with a larger difference at the edge. Centrifugal contours are nearly straight-line contours, while the improved solution contours are arc. Vertical displacement has a relatively small numerical value and its regularity is not obvious. The shape and distribution position of centrifugal contours and improved contours are similar. The difference also shows an increasing trend from the center to the edge, and the improved solution contours shift to the right a little and has a wider range. One of the important reasons is that sand soil is chosen in the centrifugal test but part of Loganathan's solution is fitted with clay data and is only applicable to clay soil. The current centrifugal tests are mostly based on sand. However, suitable clay centrifugal tests are hard to find. Considering the inevitable errors in the centrifugal test and theoretical calculations, the comparison results are acceptable. The values of Loganathan's contours are larger than the other two, because Loganathan's solution is the plane strain method; namely, it is a final stratum displacement from the excavation of an infinite tunnel. However, in the test and the improved method, the tunnel length is finite, the distance between the measuring plane and the tunnel face is much shorter, so their displacements are smaller than the final displacement.

**Author Contributions:** X.T. wrote the article; H.Z. carried out the calculations and analyzed the calculation results; G.Z. and Y.Z. processed the data; P.T. offered some useful suggestions for the preparation.

**Funding:** The study was supported by the National Natural Science Foundation of China (NSFC) under Grant No. 51508037.

**Acknowledgments:** We also highly appreciate reviewers for useful comments and editors for improving the manuscript.

**Conflicts of Interest:** The authors declare no conflict of interest.

## References

1. Zhang, H.; Chen, L.; Zhu, Y.; Zhou, Z.; Chen, S. Stress Field Distribution and Deformation Law of Large Deformation Tunnel Excavation in Soft Rock Mass. *Appl. Sci.* **2019**, *9*, 865. [CrossRef]
2. Zhang, Z.Q.; Zhang, H.; Tan, Y.J.; Yang, H.Y. Natural wind utilization in the vertical shaft of a super—Long highway tunnel and its energy saving effect. *Build. Environ.* **2018**, *145*, 140–152. [CrossRef]
3. Zhu, Y.; Chen, L.; Zhang, H.; Zhou, Z.; Chen, S. Physical and Mechanical Characteristics of Soft Rock Tunnel and the Effect of Excavation on Supporting Structure. *Appl. Sci.* **2019**, *9*, 1517. [CrossRef]
4. Zhang, H.; Chen, L.; Chen, S.G.; Sun, J.C.; Yang, J.S. The spatiotemporal distribution law of microseismic events and rockburst characteristics of the deeply buried tunnel group. *Energies* **2018**, *11*, 3257. [CrossRef]
5. Tang, F.; Li, L.J.; Dong, M.S.; Wang, Q.; Mei, F.Z.; Hu, L.H. Characterization of buoyant flow stratification behaviors by Richardson (Froude) number in a tunnel fire with complex combination of longitudinal ventilation and ceiling extraction. *Appl. Therm. Eng.* **2017**, *110*, 1021–1028. [CrossRef]
6. Mei, F.Z.; Tang, F.; Ling, X.; Yu, J.S. Evolution characteristics of fire smoke layer thickness in a mechanical ventilation tunnel with multiple point extraction. *Appl. Therm. Eng.* **2017**, *111*, 248–256. [CrossRef]
7. Sun, J.; Zhang, H.; Huang, M.; Chen, Q.; Chen, S. Reasonable Paths of Construction Ventilation for Large—Scale Underground Cavern Groups in Winter and Summer. *Sustainability* **2018**, *10*, 3768. [CrossRef]
8. Lin, F.; Chen, S.G.; Zhang, H. Study on type selection of shield equipment in different geological conditions. In *Green Building, Materials and Civil Engineering*; CRC Press: Boca Raton, FL, USA, 2014; pp. 391–394.
9. Zhang, H.; Chen, S.G.; Deng, X.F. Analysis on Influence of Shield Tunneling on Ground and Bridge Pile. *Chin. J. Undergr. Space Eng.* **2011**, *7*, 552–557.
10. Zhang, H.; Chen, S.G.; Tan, X.R. Research on the Mechanical Behavior of a Segment of a Shield Tunnel Adjacent to a Pile Foundation. *Mod. Tunn. Technol.* **2012**, *49*, 101–107.
11. Zhou, Z.L.; Chen, S.G.; Li, Y.S. Research on Predicting and Distribution of Stratum Displacement of Double—Tube Parallel Shield Tunnel. *J. Highw. Transp. Res. Dev.* **2015**, *32*, 109–117.
12. Zhang, H.; Chen, S.G.; Deng, X.F. Analysis of the Influence of Shield Driving Parameters on Ground Settlements. *Mod. Tunn. Technol.* **2010**, *47*, 48–53.
13. Lu, D.C.; Kong, F.C.; Du, X.L.; Shen, C.P.; Gong, Q.M.; Li, P.F. A unified displacement function to analytically predict ground deformation of shallow tunnel. *Tunn. Undergr. Space Technol.* **2019**, *88*, 129–143. [CrossRef]
14. Franza, A.; Marshall, A.M. Empirical and semi—Analytical methods for evaluating tunneling—Induced ground movements in sands. *Tunn. Undergr. Space Technol.* **2019**, *88*, 47–62. [CrossRef]
15. Losacco, N.; Viggiani, G. Class A prediction of mechanised tunnelling in Rome. *Tunn. Undergr. Space Technol.* **2019**, *87*, 160–173. [CrossRef]
16. Do, N.A.; Dias, D.; Oreste, P.; Djeran, I. Three—Dimensional numerical simulation of a mechanized twin tunnels in soft ground. *Tunn. Undergr. Space Technol.* **2014**, *42*, 40–51. [CrossRef]
17. Chen, R.P.; Zhu, J.; Liu, W.; Tang, X.W. Ground movement induced by parallel EPB tunnels in silty soils. *Tunn. Undergr. Space Technol.* **2011**, *26*, 163–171. [CrossRef]
18. Xie, X.Y.; Yang, Y.B.; Ji, M. Analysis of ground surface settlement induced by the construction of a large—Diameter shield—Driven tunnel in Shanghai, China. *Tunn. Undergr. Space Technol.* **2016**, *51*, 120–132. [CrossRef]
19. Peck, R.B. Deep excavations and tunneling in soft ground. State of the Art Report. In Proceedings of the 7th International Conference on Soil Mechanics and Foundation Engineering, Mexico City, Mexico, 5 August 1969; pp. 225–290.
20. Attewell, P.B. *Ground Movements Caused by Tunneling in Soil. Large Ground Movements and Structures*; Pentech Press: London, UK, 1978; pp. 812–948.

21. New, B.M.; O'Reilly, M.P. Tunnelling induced ground movements: Predicting their magnitude and effects. In *4th Int. Conf. Ground Movements and Structures*; Pentech Press: Cardiff, Wales, 1991; pp. 671–697.
22. Mair, R.J.; Gunn, M.J.; O'Reilly, M.P. Ground movement around shallow tunnels in soft clay. *Tunn. Tunn. Int.* **1982**, *14*, 45–48.
23. Mair, R.J.; Taylor, R.N.; Bracegirdle, A. Subsurface settlement profiles above tunnels in clays. *Geotechnique* **1993**, *43*, 315–320. [CrossRef]
24. Meguid, M.A.; Saada, O.; Nunes, M.A.; Mattar, J. Physical modeling of tunnels in soft ground: a review. *Tunn. Undergr. Space Technol.* **2008**, *23*, 185–198. [CrossRef]
25. Klar, A.; Vorster, T.E.B.; Soga, K.; Mair, R.J. Soil—Pipe interaction due to tunnelling: Comparison between Winkler and elastic continuum solutions. *Geotechnique* **2005**, *55*, 461–466. [CrossRef]
26. Celestino, T.B.; Gomes, R.A.M.; Bortolucci, A.A. Errors in group distortions due to settlement trough adjustment. *Tunn. Undergr. Space Technol.* **2000**, *15*, 97–100. [CrossRef]
27. Vorster, T.E.B.; Klar, A.; Soga, K.; Mair, R.J. Estimating the effects of tunneling on existing pipelines. *J. Geotech. Geoenviron. Eng.* **2005**, *131*, 1399–1410. [CrossRef]
28. O'Reilly, M.P.; New, B.M. Settlement above tunnels in the United Kingdom their magnitude and prediction. In Proceedings of the Tunneling' 82 Symposium, London, UK, 7–11 June 1982; pp. 173–181.
29. Osman, A.S.; Bolton, M.D.; Mair, R.J. Predicting 2D ground movement around tunnels in undrained clay. *Geotechnique* **2006**, *56*, 597–604. [CrossRef]
30. Pinto, F.; Whittle, A.J. Ground Movements due to Shallow Tunnels in Soft Ground. I: Analytical solutions. *J. Geotech. Geoenviron. Eng.* **2013**, *140*, 04013040. [CrossRef]
31. Litwinisyn, J. The theories and model research of movements of ground. In *Proceedings of the European Congress on Ground Movement*; University of Leeds: Leeds, UK, 1957.
32. Liu, B.C.; Zhang, J.S. Stochastic Method for Ground Subsidence Due to Near Surface Excavation. *Chin. J. Rock Mech. Eng.* **1995**, *14*, 289–296.
33. Yang, J.S.; Liu, B.C. Ground Surface Movement and Deformation Due to Tunnel Construction by Squeezing Shield. *Rock Soil Mech.* **1998**, *19*, 10–13.
34. Shi, C.H.; Cao, C.Y.; Lei, M.F. An Analysis of the Ground Deformation Caused by Shield Tunnel Construction Combining an Elastic Half—Space Model and Stochastic Medium Theory. *Korean Soc. Civ. Eng.* **2017**, *21*, 1933–1944. [CrossRef]
35. Sagaseta, C. Analysis of undrained soil deformation due to ground loss. *Geotechnique* **1987**, *37*, 301–320. [CrossRef]
36. Lee, K.M.; Rowe, R.K. Effects of undrained strength anisotropy on surface subsidences induced by the construction of shallow tunnels. *Can. Geotech. J.* **1989**, *26*, 279–291. [CrossRef]
37. Rowe, R.K.; Kack, G.J. A theoretical examination of the settlements induced by tunneling: Four case histories. *Can. Geotech. J.* **1983**, *20*, 299–314. [CrossRef]
38. Wei, G. Establishment of uniform ground movement model for shield tunnels. *Chin. J. Geotech. Eng.* **2007**, *29*, 554–559.
39. Jiang, X.L.; Zhao, Z.M. Application of image method in calculating tunneling induced soil displacement. *J. Harbin Inst. Technol.* **2005**, *37*, 801–803.
40. González, C.; Sagaseta, C. Patterns of soil deformations around tunnels. Application to the extension of Madrid Metro. *Comput. Geotech.* **2001**, *28*, 445–468. [CrossRef]
41. Zhang, Z.G.; Huang, M.S. Geotechnical influence on existing subway tunnels induced by multiline tunneling in Shanghai soft soil. *Comput. Geotech.* **2014**, *56*, 121–132. [CrossRef]
42. Randolph, M.F.; Wroth, C.P. Analysis of Deformation of Vertically Loaded Pile. *J. Geotech Geoenviron. Eng.* **1978**, *104*, 14262.
43. Park, K.H. Analytieal solution for tunneling—Induced ground movement in clays. *Tunn. Undergr. Space Technol.* **2005**, *20*, 249–261. [CrossRef]
44. Loganathan, N.; Poulos, H.G. Analytical prediction for tunneling–induced ground movements in clays. *J. Geotech. Geoenviron. Eng.* **1998**, *124*, 846–856. [CrossRef]
45. Jiang, X.L.; Zhao, Z.M.; Li, Y. Analysis and calculation of surface and subsurface settlement trough profiles due to tunneling. *Rock Soil Mech.* **2004**, *25*, 1542–1544.

46. Zhu, Y.M.; Chen, L.; Zhang, H.; Tu, P.; Chen, S.G. Quantitative Analysis of Soil Displacement Induced by Ground Loss and Shield Machine Mechanical Effect in Metro Tunnel Construction. *Appl. Sci.* **2019**, *9*, 3028. [CrossRef]
47. Cheng, C.Y.; Dasari, G.R.; Chow, Y.K.; Leung, C.F. Finite element analysis of tunnel—Soil—Pile interaction using displacement controlled model. *Tunn. Undergr. Space Technol.* **2007**, *22*, 450–466. [CrossRef]
48. Yin, M.; Jiang, H.; Jiang, Y.; Sun, Z.; Wu, Q. Effect of the excavation clearance of an under—Crossing shield tunnel on existing shield tunnels. *Tunn. Undergr. Space Technol.* **2018**, *78*, 245–258. [CrossRef]
49. Zhang, Z.G.; Zhang, M.X.; Jiang, Y.J.; Bai, Q.; Zhao, Q.H. Analytical prediction for ground movements and liner internal forces induced by shallow tunnels considering non-uniform convergence pattern and ground-liner interaction mechanism. *Soils Found.* **2017**, *57*, 211–226. [CrossRef]
50. Wang, H.N.; Zeng, G.S.; Jiang, M.J. Analytical stress and displacement around non-circular tunnels in semi-infinite ground. *Appl. Math. Model.* **2018**, *6*, 303–328. [CrossRef]
51. Xie, X.Y.; Wang, Q.; Shahrour, I.; Li, J.; Zhou, B. A real-time interaction platform for settlement control during shield tunnelling construction. *Autom. Constr.* **2018**, *94*, 154–167. [CrossRef]
52. Stevens, J. Unified soil classification system. *Civ. Eng.* **1982**, *52*, 61–62.
53. Frank, R. *Designers' Guide to EN 1997-1 Eurocode 7: Geotechnical Design-General Rules*; Thomas Telford: London, UK, 2004.
54. Hight, D.W. Laboratory testing: Assessing BS 5930. *Geol. Soc. Lond. Eng. Geol. Spec. Publ.* **1986**, *2*, 43–52. [CrossRef]
55. Kim, C.Y.; Bae, G.J.; Hong, S.W.; Park, C.H.; Moon, H.K.; Shin, H.S. Neural network based prediction of ground surface settlements due to tunnelling. *Comput. Geotech.* **2001**, *28*, 517–547. [CrossRef]
56. Tsekouras, G.J.; Koukoulis, J.; Mastorakis, N.E. An optimized neural network for predicting settlements during tunneling excavation. *WSEAS Trans. Syst.* **2010**, *9*, 1153–1167.
57. Miliziano, S.; de Lillis, A. Predicted and observed settlements induced by the mechanized tunnel excavation of metro line C near S. Giovanni station in Rome. *Tunn. Undergr. Space Technol.* **2019**, *86*, 236–246. [CrossRef]
58. Fargnoli, V.; Boldini, D.; Amorosi, A. Twin tunnel excavation in coarse grained soils: Observations and numerical back-predictions under free field conditions and in presence of a surface structure. *Tunn. Undergr. Space Technol.* **2015**, *49*, 454–469. [CrossRef]
59. Saplachidi, A.; Vougioukas, E. Comparison of Actual Ground Settlements in Tunneling Excavation to Model Predictions. *Acta Polytech.* **2018**, *58*, 47–58. [CrossRef]
60. Suwansawat, S.; Einstein, H.H. Artificial neural networks for predicting the maximum surface settlement caused by EPB shield tunneling. *Tunn. Undergr. Space Technol.* **2006**, *21*, 133–150. [CrossRef]
61. Ahangari, K.; Moeinossadat, S.R.; Behnia, D. Estimation of tunneling—Induced settlement by modern intelligent methods. *Soils Found.* **2015**, *55*, 737–748. [CrossRef]
62. Selby, A.R. Tunnelling in soils-ground movements, and damage to buildings in Workington, UK. *Geotech. Geol. Eng.* **1999**, *17*, 351–371. [CrossRef]
63. Chou, W.I.; Bobet, A. Predictions of ground deformations in shallow tunnels in clay. *Tunn. Undergr. Space Technol.* **2002**, *17*, 3–19. [CrossRef]
64. Ocak, I. Control of surface settlements with umbrella arch method in second stage excavations of Istanbul Metro. *Tunn. Undergr. Space Technol.* **2008**, *23*, 674–681. [CrossRef]
65. Darabi, A.; Ahangari, K.; Noorzad, A.; Arab, A. Subsidence estimation utilizing various approaches—A case study: Tehran No. 3 subway line. *Tunn. Undergr. Space Technol.* **2002**, *31*, 117–127. [CrossRef]
66. Zhu, C.H.; Li, N.; Liu, H.X.; Zhang, Z.Q. Analysis of ground settlement induced by workmanship of shield tunnelling. *Rock Soil Mech.* **2011**, *1*, 158–164.
67. Wei, G. *Theoretical Study on Behavior of Soil and Structure during Pipe Jacking Construction*; Hangzhou Zhejiang University: Hangzhou, China, 2005.
68. Mair, R.J.; Taylor, R.N. Theme lecture: Bored tunnelling in the urban environment. In Proceedings of the Fourteenth International Conference on Soil Mechanics and Foundation Engineering, Hamburg, Germany, 6–12 September 1997; pp. 2353–2385.
69. Rowe, R.K.; Lo, K.Y.; Kack, G.J. A method of estimating surface settlement above tunnels constructed in soft ground. *Can. Geotech. J.* **1983**, *20*, 11–22. [CrossRef]
70. Lee, K.M.; Rowe, R.K.; Lo, K.Y. Subsidence owing to tunneling I: Estimating the gap parameter. *Can. Geotech. J.* **1992**, *29*, 929–940. [CrossRef]

71. Verruijit, A.; Booker, J.R. Surface settlements due to deformation of a tunnel in an elastic half plane. *Geotechnique* **1996**, *46*, 753–756. [CrossRef]
72. Verruijt, A. Deformations of an elastic half plane with a circular cavity. *Int. J. Solids Struct.* **1998**, *35*, 2795–2804. [CrossRef]
73. Marshall, A.M. *Tunneling in Sand and Its Effect on Pipelines and Pipe*; University of Cambridge: London, UK, 2009.
74. Vorster, T.E.B. The Effects of Tunnelling on Buried Pipes. Ph.D. Thesis, Cambridge University, Cambridge, UK, 2005.
75. Lee, S. The Effects of Compensation Injections on Tunnels. Ph.D. Thesis, Cambridge University, Cambridge, UK, 2001.
76. Tan, W.L.; Ranjith, P.G. Parameters and considerations in soft ground tunneling. *Electron. J. Geotech. Eng.* **2003**, *8*, 1.

applied sciences

Article

# A Precise Prediction of Tunnel Deformation Caused by Circular Foundation Pit Excavation

Huasheng Sun [1,*], Lingwei Wang [1], Shenwei Chen [1,2], Hengwei Deng [1] and Jihua Zhang [1]

[1] Department of Civil Engineering, Huaiyin Institute of Technology, Huai'an 223001, China; wanglw1996@hotmail.com (L.W.); huzcsw@hotmail.com (S.C.); dhw1999@hotmail.com (H.D.); zhangjh84@hyit.edu.cn (J.Z.)
[2] Department of Civil Engineering, Nanjing Tech University, Nanjing 211816, China
* Correspondence: sunhuasheng@hyit.edu.cn or sunhuadasheng@126.com; Tel.: +86-18262811276

Received: 29 April 2019; Accepted: 30 May 2019; Published: 2 June 2019

**Abstract:** In comparison with tetragonal retaining structures, circular retaining structures have an advantage in terms of controlling the deformation caused by foundation excavation, and are a reasonable choice in engineering practice. Many results have been obtained regarding the effect of tetragonal excavation on the deformation of an adjacent tunnel. Nevertheless, a sufficient understanding of the circular excavation's effect on the deformation of an adjacent tunnel is currently lacking. Therefore, this study focused on the problem of precise predicting tunnel deformation below a circular excavation. A numerical model was established to calculate the tunnel deformation caused by the circular excavation. An advanced nonlinear constitutive model, known as a hypoplasticity model, which can capture path-dependent and strain-dependent soil stiffness even at small strains, was adopted. The models and their associated parameters were calibrated by centrifuge test results reported in the literature. The deformation mechanism was revealed, and the calculated results were compared with those obtained with a square excavation and the same excavation amount. The differences between the deformations caused by these two types of excavation shapes were analyzed. It was found that under equal excavation area conditions, the excavation-induced deformations of the metro tunnel below a circular excavation were approximately 1.18–1.22 times greater than those below a square excavation. The maximum tunnel tensile bending strain caused by the circular excavation was 32% smaller than that caused by the square excavation. By comparing with the measured results, it is proved that the proposed numerical method can provide effective reference for engineers to analyze soil-structure problems.

**Keywords:** advanced model; precise prediction; circular foundation pit; tunnel deformation

## 1. Introduction

With urban development, a large number of excavation works have been conducted. Stress relief caused by basement excavation will inevitably affect surrounding subway tunnels. Many research results have been obtained with regard to the effect of tetragonal excavation on the deformation of an adjacent metro tunnel. The main contents of this study are as follows: geometry of the excavation [1,2]; relative position of tunnel and excavation [3–7]; excavation conditions and reinforcement method [8–11]; tunnel lining stiffness [4] and diameter of tunnel lining [1]; different soil constitutive models [12]; soil relative density and retaining wall stiffness [13]; influence zone [14] and calculation methods [15–17].

However, existing studies have mainly focused on analyzing the tunnel response caused by the unloading of the tetragonal basement excavation. With the rapid development of urban construction, the development and utilization of underground space is rapidly increasing and more and more foundation basements are constructed as circular foundation basements [18–21]. Unlike the traditional tetragonal rectangular basement, a circular basement can make full use of the circular arch effect, such that the force of the structure becomes more reasonable.

Existing circular basement studies have mainly focused on the soil pressure around the basement and the retaining structure's deformation caused by the excavation [22–30]. However, few studies have focused on the deformation of surrounding foundations and buildings, which is caused by the unloading of the circular excavation. Considering the many differences between circular and tetragonal basements, it is theoretically and practically significant to evaluate the effect of circular basement excavation on the deformation of the surrounding structure to reduce damages.

The objective of this study was to investigate the contribution of circular basement construction to the deformation of a nearby existing tunnel, and to reveal the tunnel deformation mechanism. Three-dimensional numerical analyses using advanced constitute model combined with accurate numerical simulation techniques were conducted based on the reported centrifuge model test. The effect of the excavation shape and the tunnel and soil responses were analyzed with the expectation that the results may provide a useful reference for engineering practice.

## 2. Introduction of a Centrifuge Model Test

A centrifuge model test, which models the interaction between a new basement excavation and an existing tunnel was carried, out as shown in Figure 1 [7]. The centrifuge acceleration of the test was 60 g. The length, width, and depth of the excavation was 300 × 300 × 150 mm, which is equal to 18 × 18 × 9 m in the prototype, respectively. The Young's modulus of the model diaphragm wall was 70 GPa, and was equal to that of the model tunnel. The model tunnel had a diameter of 100 mm, length of 1,200 mm, and thickness of 3 mm, which is equal to 6 m, 72 m, and 0.18 m in the prototype, respectively. The details of the proposed centrifuge test have been reported by Ng et al. (2013) [7]. It is also noted that the following measured results in this paper are all reported by Ng et al. (2013) [7].

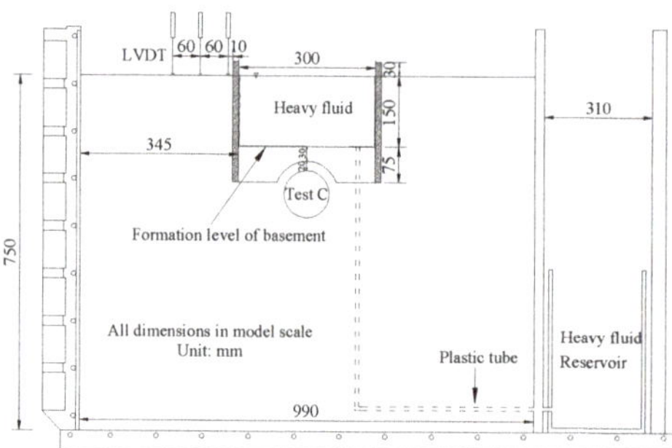


**Figure 1.** Layout of the proposed centrifuge model test [7].

## 3. Finite Element Model

### 3.1. Finite Element Mesh and Boundary Condition

In this study, the finite element software ABAQUS [31] was used. A three-dimensional finite element mesh was built as shown in Figure 2. The finite element mesh in this model had a length of 1200 mm, width of 990 mm, and depth of 750 mm. The square basement excavation (on plan) was the same as that in the centrifuge model test, with a length and width of 300 × 300 mm (equal to 18 × 18 m in the prototype). Preliminary calculations checked the influence of the mesh density (13,540–86,590 elements in a structured or unstructured mesh, respectively) and of the element type (triangle or quadrilateral elements with linear interpolation). The influence of these factors on the results has been shown to be small. However, for precise and quick convergence, 86,950 elements in a structured mesh with quadrilateral element type was adopted. Based on the principle of the excavation amount being equal to the square excavation amount recorded in the test, a circular excavation was conducted with a diameter of 339 mm (equal to 20.3 m in the prototype). The basement excavation depth was 150 mm, which is equal to 9 m in the prototype. The embedment depth of the diaphragm wall was 225 mm, which is equal to 13.5 m in the prototype. The sand and the diaphragm wall were simulated by an eight-node brick element, and the tunnel was simulated by a four-node shell element. To consider the friction boundary condition, pin supports were adopted both at the vertical sides and at the base of the model. Interface elements were adopted at the soil-diaphragm wall and at the soil-tunnel interfaces. Each interface element used is described by zero-thickness slip element, assigned with the Coulomb friction law. The friction coefficient ($\mu$) and limiting relative displacement ($\gamma_{\text{lim}}$) at which slippage occurs are controlled by two input parameters for each slip element. The interface friction coefficient, $\mu$, is derived from $\mu = \tan\delta$, where $\delta$ is the interface friction angle, which is taken as 20° (i.e., 2/3 of the critical friction angle of soil). The limiting displacement of 5 mm is assumed to achieve full mobilization of the interface friction. It is noted that the initial compaction due to air pressure was not taken into account.

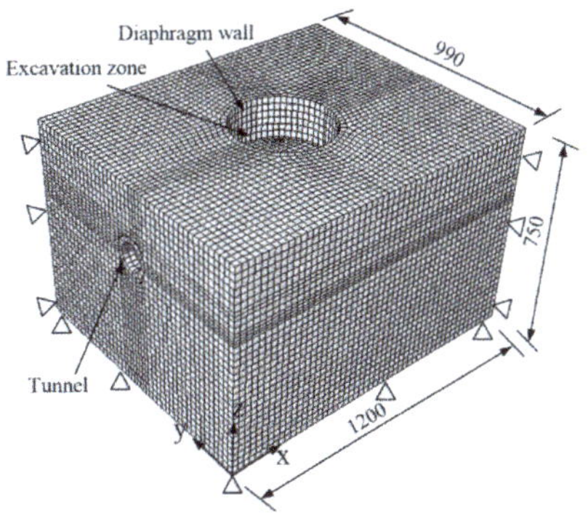


**Figure 2.** Finite element model (in model scale, unit: mm).

### 3.2. Constitutive Model and Model Parameters

In the references paper [12], the ability of three different soil constitutive models to predict a tunnel's response to basement excavation has been evaluated. It is found that the hypoplastic model yielded the best predictions of tunnel heave among the adopted models. The nonlinearity of soil can be captured by a hypoplastic (HP) constitutive model. Various HP models have been developed in the 1990s (Kolymbas 1991; Gudehus 1996; Von Wolffersdorff 1996; Wu et al. 1996) [32–35] as well as recently (Mašín 2012 2014, 2013) [36–38]. The model proposed by Von Wolffersdorff (1996) [34] was adopted to describe the behavior of Toyoura sand. This model was incorporated into the software package ABAQUS using open-source implementation which can be freely downloaded from the web (Gudehus et al. 2008) [39]. The model specifies eight material parameters ($\varphi_c$, $h_s$, $n$, $e_{d0}$, $e_{c0}$, $e_{i0}$, $\alpha$ and $\beta$). Niemunis and Herle (1997) [40] improved the model for predictions of small-strain stiffness and the recent stress history, leading to five additional parameters ($m_T$, $m_R$, $R$, $\beta_r$ and $\chi$). See Table 1 and the literature mentioned above for their physical meaning.

**Table 1.** Soil parameters adopted in the hypoplastic model.

| Soil Parameters | Values |
|---|---|
| Critical state friction angle [(1)], $\varphi_c$ | 30° |
| Granulates hardness [(1)], $h_s$ | 2.6 GPa |
| Exponent $n$ [(1)], $n$ | 0.27 |
| Minimum void ratio at zero pressure [(1)], $e_{d0}$ | 0.61 |
| Critical void ratio at zero pressure [(1)], $e_{c0}$ | 0.98 |
| Maximum void ratio at zero pressure [(1)], $e_{i0}$ | 1.10 |
| Exponent $\alpha$ [(2)], $\alpha$ | 0.14 |
| Exponent $\beta$ [(2)], $\beta$ | 3.0 |
| Parameter controlling the initial shear modulus upon a 180° strain path reversal and in the initial loading [(2)], $m_R$ | 5.5 |
| Parameter controlling the initial shear modulus upon a 90° strain path reversal [(2)], $m_T$ | 2.75 |
| Size of elastic range [(2)], $R$ | $3 \times 10^{-5}$ |
| Parameter controlling rate of degradation of stiffness with strain [(2)], $\beta_r$ | 0.08 |
| Parameter controlling rate of degradation of stiffness with strain [(2)], $\chi$ | 1.0 |
| Coefficient of at-rest earth pressure, $K_0$ | 0.5 |
| Dry density, $\rho_d$ | 1542 kg/m$^3$ |
| Void ratio, $e$ | 0.72 |

Note: (1) Herle and Gudehus, 1999 [41]. (2) Obtained by fitting test results from Maeda and Miura (1999) [42] and Iwasaki et al. (1978) [43].

Six parameters of Toyoura sand ($\varphi_c$, $h_s$, $n$, $e_{d0}$, $e_{c0}$ and $e_{i0}$) were obtained from Herle and Gudehus (1999) [41], while the triaxial test results reported by Maeda and Miura (1999) [42] were used to calibrate the parameters of $\alpha$ and $\beta$. Five parameters ($m_T$, $m_R$, $R$, $\beta_r$ and $\chi$) of the intergranular strain can be calibrated from the stiffness degradation curve of Toyoura sand (Iwasaki et al. 1978) [43], as shown in Figure 3. The void ratio of soil was considered as a state variable in the HP model. The parameter selection list is presented in Table 1. The parameter selection procedure and values have been reported by Ng et al. (2015) [12]. The tunnel and diaphragm wall can be considered as an elastic material. Their elastic modulus and Poisson's ratio were 70 GPa and 0.2, respectively.

### 3.3. Numerical Modelling Procedures

The gradual unloading method was adopted to simulate the excavation. Figure 3 shows the flow chart of specific simulation process and as follows.

1. Establish the initial stress conditions using $K_0 = 0.5$. Apply the same amounts of vertical and horizontal pressure as that in the centrifuge test to the formation level and diaphragm wall, respectively.
2. Incrementally increase the gravitational acceleration of the whole model from 1 g to 60 g in four steps, i.e., from 1 g, to 15 g, to 30 g, to 45 g, and finally to 60 g. Simultaneously, apply pressure to the formation level and the wall.
3. Decrease the amounts of vertical and horizontal pressure gradually in each excavation stage to simulate excavation until a depth of 9 m is reached.

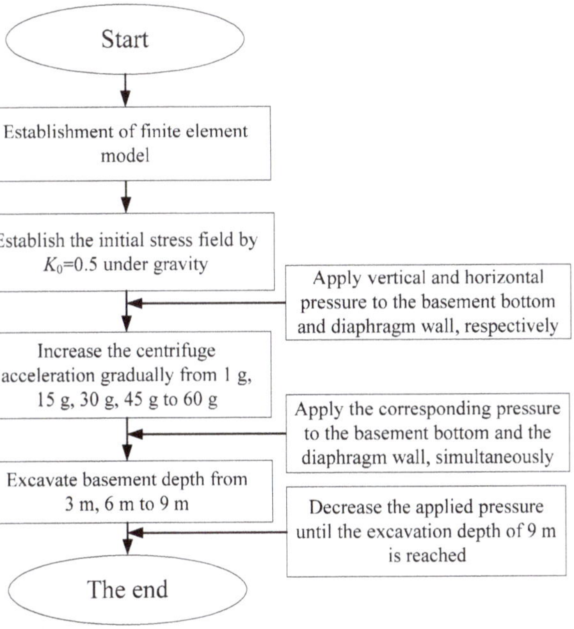


**Figure 3.** Flow chart of the numerical modelling procedures.

## 4. Result Interpretation

### 4.1. Tunnel Heave in Longitudinal Direction

Figure 4 shows the measured tunnel heave computed by numerical analyses for both the square and the circular basement excavations. It can be seen that the computed tunnel heave was generally in agreement with the measured magnitude and distribution values. The tunnel heave in the existing tunnel was caused by stress relief, as shown in Figure 5. With the increase of the basement excavation depth, the tunnel heave increased. After the basement excavation, the measured tunnel heave reached 0.07% $H_e$ ($H_e$ is the basement depth) at a distance of 0.4 $H_e$ from the basement center, and reduced to zero at a distance of 2.4 $H_e$ from the basement center. The computed maximum value of the tunnel heave was 0.084% $H_e$ and 0.099% $H_e$ for the square and circular basement excavations, respectively, with a similar trend distribution. By comparing the computed tunnel heave caused by a different basement shape, it was found that the maximum tunnel heave caused by the unloading of the circular excavation was 1.18 times greater than that of the square excavation with the same excavation amount. According to the provisions of the Land Transport Authority [44], the allowable maximum tunnel heave value is 15 mm (i.e., 0.170% $H_e$, as shown in Figure 4). The computed tunnel heave ranged in the standard allowable value and also in the measured value. From the above analysis, it was determined that the tunnel heave caused by the circular basement excavation was 18% larger than that caused by the square basement excavation. Additionally, if a circular basement is adopted, and assuming that the tunnel heave caused by the square basement is in a critical state, it is essential to take corresponding reinforcement measures.

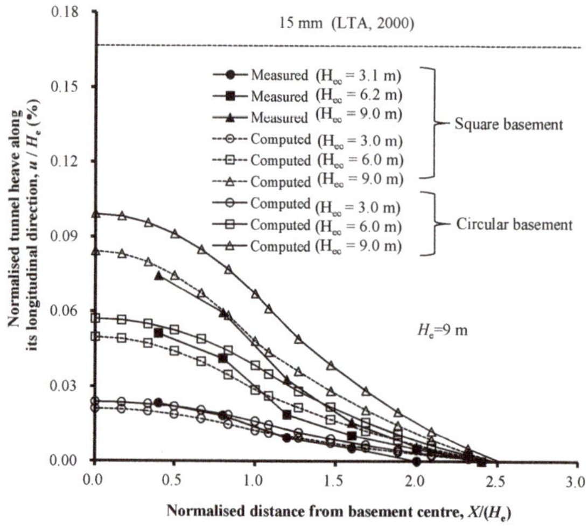


**Figure 4.** Measured and computed tunnel heave.

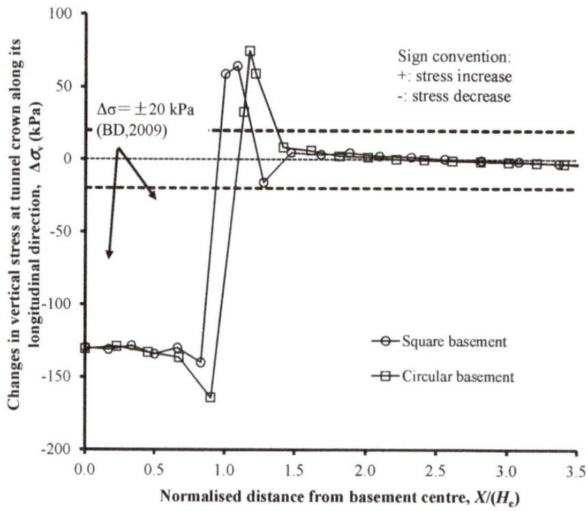


**Figure 5.** Vertical stress changes in soil element around tunnel crown.

## 4.2. Stress Distribution around the Tunnel Lining

To explain the reason for the occurrence of tunnel heave, Figure 5 shows the stress distribution in the soil element around the tunnel crown. The stress increases are shown as positive values, while the stress decreases are shown as negative values. The basement excavation with two different shapes predicted almost similar vertical stress patterns for the soil elements around the tunnel crown. A great decrease in the vertical stress beneath the basement was observed after the excavation. Additionally, it was observed that the vertical soil stress was uniform beneath the basement, and increased greatly just beneath the diaphragm wall. The vertical stress increased by 64 kPa and 75 kPa for the square and circular basement, respectively. The largest stress change for the soil element was caused by the circular basement around the tunnel crown and was 17% larger than that of the square basement. Thus, the tunnel heave caused by the circular basement excavation was larger than that caused by the square basement excavation, as shown in Figure 4. At a distance 0.6 $H_e$ from the diaphragm wall location, the stress change in the soil element was less than 20 kPa. Based on the Building department of the government of Hong Kong Special Administrative Region, HKSAR [45], the stress change in this case did not exceed the proposed limit value (i.e., ±20 kPa).

### 4.3. Tunnel Diameter Change

Figure 6 shows the tunnel diameter ($D$) change varying with $H_{ec}/C$ (defined as the unloading ratio), where $H_{ec}$ is the current excavation depth and $C$ is the tunnel cover depth. The positive value represents the elongation, while the negative value represents compression. The tunnel lining caused by both basement excavation types was elongated vertically and compressed horizontally. It is clear that with the increase of the basement excavation depth, the elongation ($\Delta D_V$) and compression ($\Delta D_H$) of the tunnel lining increased. The square and circular basement resulted in a maximum change in the tunnel diameter by 0.09% $D$ and 0.11% $D$, respectively. The maximum tunnel diameter change caused by the circular basement excavation was 22% larger than that caused by the square basement.

To better understand the tunnel diameter change for the two different types of basement excavation, the earth pressure change around the tunnel is shown in Figure 7, where it can be seen that the maximum earth pressure change occurred in the soil around the tunnel crown, while a smaller change occurred at the springlines, and little change occurred at the invert. Because the earth pressure reduction was larger at the crown than at the springlines, the tunnel lining was elongated vertically and compressed horizontally as shown in Figure 6. By comparing the two types of basement excavation, the earth pressure change at the tunnel crown and invert was almost the same because of the same excavation amount assumed by the model. However, owing to the effect of arching, the earth pressure change for the other soil elements of the square basement was smaller than that in the circular basement excavation. Thus, the tunnel diameter change caused by the circular basement excavation was larger than that caused by the square basement, as shown in Figure 6. Based on the recommendations of the British Tunnelling Society [46], the maximum tunnel lining change (($\Delta D_V$ + $\Delta D_H$)/$D$) should not exceed 2%. Thus, it was determined that the maximum deformation of the tunnel (i.e., 0.16% $D$) was smaller than the proposed allowable limit.

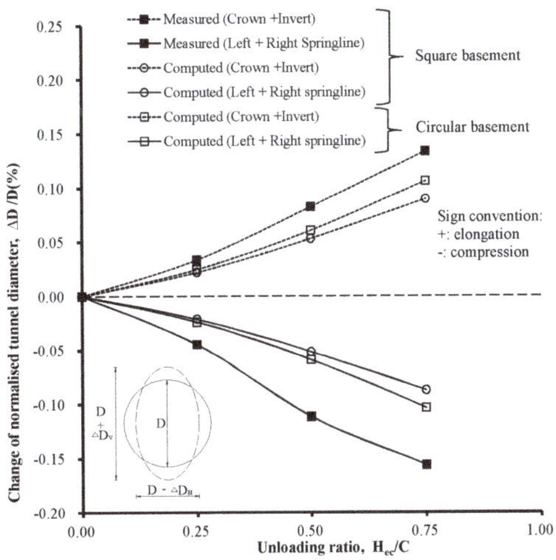


**Figure 6.** Comparison of tunnel diameter change.

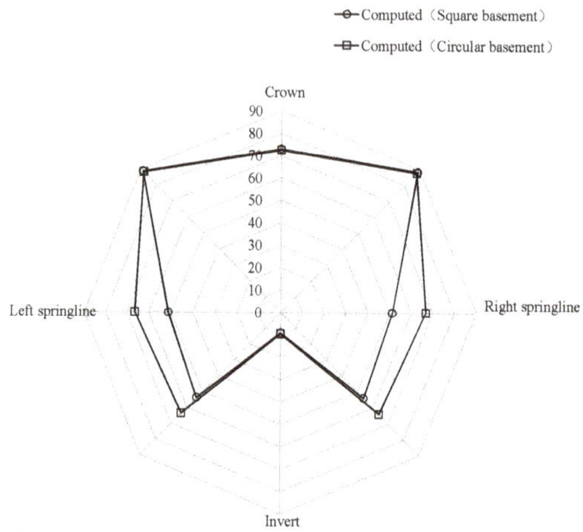


**Figure 7.** Earth pressure change around the tunnel lining (absolute value).

### 4.4. Bending Strain at the Tunnel Lining

The bending strains computed at the tunnel lining in the transverse direction were compared with the measured values, as shown in Figure 8a. The positive value represents the tensile strain, and the negative value represents the compressive strain. In comparison with the measured results, the computed results for both basement types exhibited similar trends. Moreover, the symmetrical bending strain distribution with regard to the tunnel centerline could be determined. The compressive strain occurred at the tunnel springlines, while the tensile strain occurred at the other tunnel elements. In the transverse direction, the maximum tensile bending strain was 91 με and 108 με for the square and circular basement excavation, respectively. There was only a small change in the transverse direction for these two excavations types, not only in the magnitude, but also in the distribution. Thus it was concluded that, in the transverse direction, the basement shape had little influence on the tunnel's bending strain. Based on the American Concrete Institute [35], the maximum tensile strain was 150 με. The computed bending strains did not exceed the given allowable limit nor the measured value, assuming that zero bending strain occurred before the basement construction.

Figure 8b shows the bending strains computed after the excavation in the longitudinal direction in comparison with the measured values. The positive sign denotes the tensile strain, while the negative sign denotes the compressive strain occurring at the tunnel lining crown. Similar trends were observed in the measured and computed results. With regard to the basement center, it was found that the distribution of the bending strain was symmetrical. For the square and circular basement, the maximum tensile strain computed by the numerical method was 86 με and 65 με, respectively. The computed maximum tunnel lining bending strain in the longitudinal direction, which was caused by the circular excavation, was 32% smaller than that caused by the square excavation. Based on ACI [47], cracks may occur in the tunnel lining assuming that the tensile strain exceeds 85 με before the circular basement excavation.

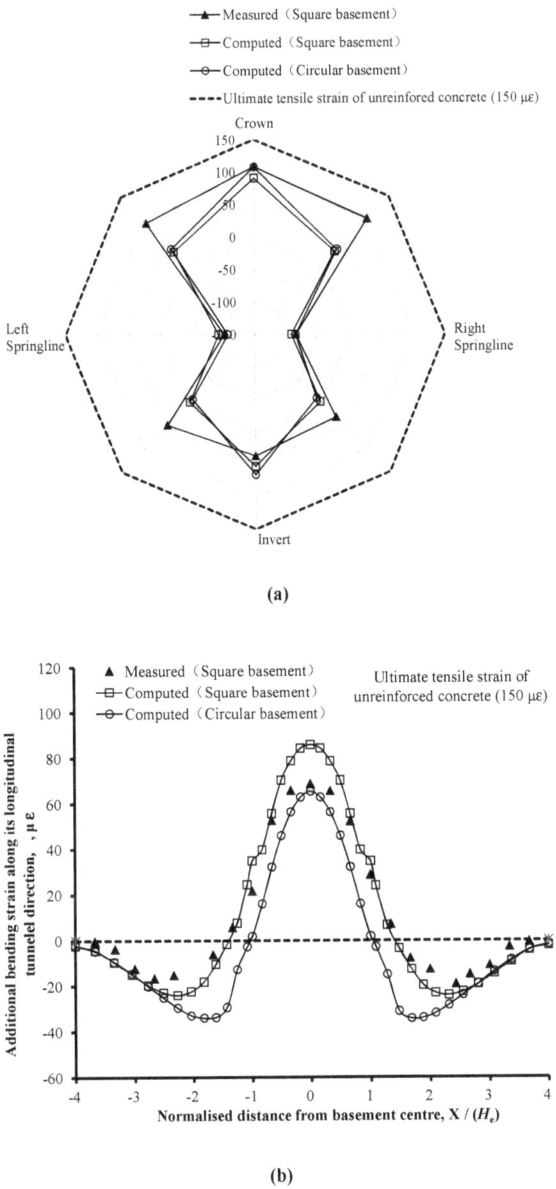


**Figure 8.** Additional bending strains in tunnel lining along the (**a**) transverse and (**b**) longitudinal direction.

## 5. Conclusions

This paper studies the tunnel deformation caused by circular excavations; the problem is tackled numerically by means an advanced nonlinear hypoplasticity constitutive model, which is able to capture the path-dependent and strain-dependent soil stiffness, even at small strains. The numerical analysis is accompanied by an experimental calibration of the parameters based on centrifuge tests. The effect of the excavation shape and the tunnel and soil responses were analyzed. The precise prediction method is useful for practical applications and can represent a key reference the design of tunnel structures, accounting for the soil-structural interaction. The comparisons of the predicted results can be regarded as the verification of different design analyses, carried out by practicing engineers. Based on the comparisons between measured and computed results, the following conclusions may be drawn:

(a) Tunnel heave occurred in the existing tunnel lining, owing to stress relief above the existing tunnel. Thus, the tunnel heave caused by the circular basement excavation was 18% larger than that caused by the square basement.
(b) The tunnel lining caused by both basement excavation types was elongated vertically and compressed horizontally with the basement excavation depth. The maximum tunnel diameter change caused by the circular basement excavation was 22% larger than that caused by the square basement excavation.
(c) In the transverse direction, the compressive strains were recorded at the tunnel springlines, while the tensile strains arose at the other tunnel elements. The basement shape had little influence on the tunnel bending strain. In the tunnel's longitudinal direction, the symmetrical bending strain distribution could be determined with regard to the basement excavation centerline. The computed maximum tunnel bending strain caused by the circular excavation was 32% smaller than that caused by the square excavation.
(d) Under identical excavation area conditions, the extent of the influence of the stress relief caused by the circular excavation was wider and deeper than that caused by the square excavation. This implies that controlling the deformation of a metro tunnel under excavation depends mainly on the length of the metro tunnel passing through the excavation area, rather than on the excavation shape.

**Data Availability Statement:** The computed data used to support the findings of this study are included within the article. Previously reported measured data were used to support this study and are available at [https://doi.org/10.1139/cgj-2012-0423, https://doi.org/10.1139/cgj-2014-0361]. These prior studies (and datasets) are cited at relevant places within the text as references [7,12].

**Author Contributions:** Conceptualization, H.S.; methodology, H.S.; software, H.S. and S.C.; validation, H.S. and L.W.; formal analysis, H.S. and L.W.; investigation, H.S. and H.D.; resources, H.S. and S.C.; data curation, H.S. and H.D.; writing—original draft preparation, H.S. and L.W.; writing—review and editing, H.S. and J.Z.; visualization, H.S.; supervision, H.S.; project administration, H.S.; funding acquisition, H.S.

**Funding:** This research was funded by the National Natural Science Foundation of China, grant number 51708245, the Natural Science Foundation of Jiangsu Province, grant number BK20160426, and the Natural Science Foundation for Colleges and Universities in Jiangsu Province (17KJB620002, 17KJB130003, 17KJA560001).

**Acknowledgments:** We thank the anonymous reviewers for their valuable comments and suggestions. We thank Liwen Bianji, Edanz Editing China (www.liwenbianji.cn/ac), for editing the English text of a draft of this manuscript.

**Conflicts of Interest:** The authors declare that there are no conflicts of interest regarding the publication of this paper.

## References

1. Huang, X.; Schweiger, H.F.; Huang, H.W. Influence of deep excavations on nearby existing tunnels. *Int. J. Geomech.* **2013**, *13*, 170–180. [CrossRef]
2. Shi, J.W.; Ng, C.W.; Chen, Y.H. Three-dimensional numerical parametric study of the influence of basement excavation on existing tunnel. *Comput. Geotech.* **2015**, *63*, 146–158. [CrossRef]
3. Doležalová, M. Tunnel complex unloaded by a deep excavation. *Comput. Geotech.* **2001**, *28*, 469–493. [CrossRef]
4. Sharma, J.S.; Hefny, A.M.; Zhao, J.; Chan, C.W. Effect of large excavation on displacement of adjacent MRT tunnels. *Tunn. Undergr. Space Technol.* **2001**, *16*, 93–98. [CrossRef]
5. Zheng, G.; Wei, S.W. Numerical analysis of influence of overlying pit excavation on existing tunnels. *J. Cent. South Univ. Technol.* **2008**, *15*, 69–75. [CrossRef]
6. Huang, X.; Huang, H.W.; Zhang, D.M. Centrifuge modeling of deep excavation over existing tunnels. *Geotech. Eng.* **2014**, *167*, 3–18. [CrossRef]

7. Ng, C.W.; Shi, J.; Hong, Y. Three-dimensional centrifuge modelling of basement excavation effects on an existing tunnel in dry sand. *Can. Geotech. J.* **2013**, *50*, 874–888. [CrossRef]
8. Hu, Z.F.; Yue, Z.Q.; Zhou, J.; Tham, L.G. Design and construction of a deep excavation in soft soils adjacent to the Shanghai Metro tunnels. *Can. Geotech. J.* **2003**, *40*, 933–948. [CrossRef]
9. Gao, G.; Gao, M.; Yang, C. Influence of deep excavation on deformation of operating metro tunnels and countermeasures. *Chin. J. Geotech. Eng.* **2010**, *32*, 453–459.
10. Zheng, G.; Du, Y.; Diao, Y. Optimization Analysis of Efficiency of Isolation Piles Incontrolling the Deformation of Existing Tunnels Adjacent to Deep Excavation. *Chin. J. Rock Mech. Eng.* **2015**, *34*, 3499–3509.
11. Feng, S.J.; Gao, G.Y.; AI, H.T. Deformation analysis of group excavation adjacent to subway tunnels. *Chin. J. Geotech. Eng.* **2008**, *30*, 112–117.
12. Ng, C.W.; Sun, H.S.; Lei, G.H.; Shi, J.W.; Mašín, D. Ability of three different soil constitutive models to predict a tunnel's response to basement excavation. *Can. Geotech. J.* **2015**, *52*, 1685–1698. [CrossRef]
13. Ng, C.W.; Shi, J.W.; David, M.; Sun, H.S.; Lei, G.H. Influence of sand density and retaining wall stiffness on three-dimensional responses of tunnel to basement excavation. *Can. Geotech. J.* **2015**, *52*, 1811–1829. [CrossRef]
14. Zheng, G.; Du, Y.; Diao, Y. Influenced zones for deformation of existing tunnels adjacent to excavations. *Chin. J. Geotech. Eng.* **2016**, *38*, 599–612.
15. Zhang, Z.G.; Huang, M.S.; Wang, W.D. Evaluation of deformation response for adjacent tunnels due to soil unloading in excavation engineering. *Tunn. Undergr. Space Technol.* **2013**, *38*, 244–253. [CrossRef]
16. Zhang, J.F.; Chen, J.J.; Wang, J.H. Prediction of tunnel displacement induced by adjacent excavation in soft soil. *Tunn. Undergr. Space Technol.* **2013**, *36*, 24–33. [CrossRef]
17. Shi, J.; Ng, C.W.; Chen, Y. A simplified method to estimate three-dimensional tunnel responses to basement excavation. *Tunn. Undergr. Space Technol.* **2017**, *62*, 53–63. [CrossRef]
18. Zhou, J.J.; Rao, S.L. The design and construction of large mixed foundation in west anchorage of Humen Bridge. *Bridge Constr.* **1995**, *2*, 44–47.
19. Liu, S.X. The design of circular diaphragm wall in the four phase of the project of Qinhuangdao port coal terminal dumper. *Port Eng. Technol.* **1995**, *4*, 28–35.
20. Zhu, F. Monitoring and Analysis of a Deep Small Diameter Circular Excavation. Master's Thesis, Shanghai Jiao Tong University, Shanghai, China, 2007.
21. Zhang, Z.J.; Tang, Y.; Wang, J.H. Measured deformation of a semi-circular foundation pit in Shanghai soft soils. *Chin. J. Geotech. Eng.* **2014**, *36*, 451–454.
22. Prater, E.G. An examination of some theories of earth pressure on shaft linings. *Can. Geotech. J.* **1977**, *14*, 91–106. [CrossRef]
23. Dong, X.P.; Guo, Q.H.; Zhou, S.H. Character and factor analysis of deformation of diaphragm wall in the cylindrical foundation pit. *Chin. J. Undergr. Space Eng.* **2005**, *4*, 196–199.
24. Liu, F.Q.; Wang, J.H. The active earth pressure of circular pit subjected to non-uniform surcharge loading. *J. Shanghai Jiaotong Univ.* **2006**, *40*, 2134–2136.
25. Cheng, Y.M.; Au, S.K.; Hu, Y.Y.; Wei, W.B. Active pressure for circular cut with Berezantzev's and Prater's theories, numerical modeling and field measurements. *Soils Found.* **2008**, *48*, 621–631. [CrossRef]
26. Tobar, T.; Meguid, M.A. Comparative evaluation of methods to determine the earth pressure distribution on cylindrical shafts: A review. *Tunn. Undergr. Space Technol.* **2010**, *25*, 188–197. [CrossRef]
27. Kim, K.Y.; Lee, D.S.; Cho, J. The effect of arching pressure on a vertical circular shaft. *Tunn. Undergr. Space Technol.* **2013**, *37*, 10–21. [CrossRef]
28. Liu, F.Q. Lateral earth pressures acting on circular retaining walls. *Int. J. Geomech.* **2014**, *14*, 613–624. [CrossRef]
29. Meguid, M.A.; Tran, V.D.H.; Chouinard, L.E. Discrete element and experimental investigations of the earth pressure distribution on cylindrical shafts. *Int. J. Geomech.* **2014**, *14*, 80–91.
30. Zhang, J.G.; Ren, P. Numerical simulation for constructionprocess of a very deep circular pit. In Proceedings of the International Conference on Transportation Engineering, Chengdu, China, 23–25 July 2011; pp. 1786–1791.
31. ABAQUS, Inc. *ABAQUS User's and Theory Manuals*; Version 6.17; ABAQUS, Inc.: Providence Rhode Island, RI, USA, 2017.
32. Kolymbas, D. An outline of hypoplasticity. *Arch. Appl. Mech.* **1991**, *61*, 143–151.

33. Gudehus, G. A comprehensive constitutive equation for granular materials. *Soils Found.* **1996**, *36*, 1–12. [CrossRef]
34. Von Wolffersdorff, P.A. A hypoplastic relationship for granular material with a predefined limit state surface. *Mech. Cohesive Frict. Mater.* **1996**, *1*, 251–271. [CrossRef]
35. Wu, W.; Bauer, E.; Kolymbas, D. Hypoplastic constitutive model with critical state for granular materials. *Mechanics of Materials* **1996**, *23*, 45–69. [CrossRef]
36. Mašín, D. Hypoplastic Cam-clay model. *Géotechnique* **2012**, *62*, 549–553. [CrossRef]
37. Mašín, D. Clay hypoplasticity with explicitly defined asymptotic states. *Acta Geotech.* **2013**, *8*, 481–496. [CrossRef]
38. Mašín, D. Clay hypoplasticity model including stiffness anisotropy. *Géotechnique* **2014**, *64*, 232–238. [CrossRef]
39. Gudehus, G.; Amorosi, A.; Gens, A.; Herle, I.; Kolymbas, D.; Mašín, D.; Muir Wood, D.; Niemunis, A.; Nova, R.; Pastor, M.; et al. The soilmodels.info project. *Int. J. Numer. Anal. Methods Geomech.* **2008**, *32*, 1571–1572. [CrossRef]
40. Niemunis, A.; Herle, I. Hypoplastic model for cohensionless soils with elastic strain range. *Mech. Cohesive Frict. Mater.* **1997**, *2*, 279–299. [CrossRef]
41. Herle, I.; Gudehus, G. Determination of parameters of a hypoplastic constitutive model from properties of grain assemblies. *Mech. Cohesive Frict. Mater.* **1999**, *4*, 461–486. [CrossRef]
42. Maeda, K.; Miura, K. Confining stress dependency of mechanical properties of sands. *Soils Found.* **1999**, *39*, 53–67. [CrossRef]
43. Iwasaki, T.; Tatsuoka, F.; Takagi, Y. Shear moduli of sands under cyclic torsional shear loading. *Soils Found.* **1978**, *18*, 39–50. [CrossRef]
44. LTA. *Code of Practice for Railway Protection*; Development & Building Control Department, Land Transport Authority (LTA): Singapore, 2000.
45. BD. Practice Note for Authorized Persons APP-24. In *Technical Notes for Guidance in Assessing the Effects of Civil Engineering Construction/Building Development on Railway Structures and Operations*; Building Department of the Government of HKSAR (BD): Hong Kong, China, 2009.
46. British Tunnelling Society (BTS). *Specification for Tunnelling*; Thomas Telford: London, UK, 2000.
47. American Concrete Institute. *Control of Cracking in Concrete Structures (ACI 224R-01)*; American Concrete Institute: Farmington Hills, MI, USA, 2001.

*applied sciences*

MDPI

*Article*

# Stochastic Natural Frequency Analysis of Composite Structures Based on Micro-Scale and Meso-Scale Uncertainty

**Shufeng Zhang * and Xun Chen**

Science and Technology on Integrated Logistics Support Laboratory, National University of Defense Technology, Changsha 410073, China

* Correspondence: sfzhang@nudt.edu.cn; Tel.: +86-731-8457-3396

Received: 6 June 2019; Accepted: 25 June 2019; Published: 27 June 2019

**Abstract:** Composite structure often shows undesirably significant uncertainty in its mechanical properties, which may consequently result into large stochastic variation of its natural frequency. This study provides stochastic natural frequency analysis of typical composite structures based on micro-scale (constituent-scale) and meso-scale (ply-scale) uncertainty. Uncertainty propagation across micro-scale and meso-scale is investigated. Response surface method (RSM) based on finite element modeling is employed to obtain approximate natural frequency of structures with complex shape or boundary conditions, and mean value and standard deviation of natural frequency of composite plate and cylindrical shell are derived. Differences in natural frequency statistics of composite plates and cylindrical shells derived by considering uncertainty at different scales are quantified and discussed. Significant statistical correlation between ply elastic properties and ply density is observed, and the statistical correlation is demonstrated to lay great influence on the statistics of structure natural frequency.

**Keywords:** composite; stochastic; natural frequency; uncertainty

## 1. Introduction

Fibre reinforced composite material and structures are extensively used in diverse engineering applications such as aircraft, rockets, ships, wind turbines, buildings, etc. Due to complexity in material microstructure and manufacture process, composite structure often shows undesirably large uncertainty in its mechanical performance, such as structure deflection [1,2], load capacity [3,4], buckling behaviour [5,6] and dynamic response [7]. Therefore, structure design in stochastic or probabilistic approaches is especially important for composite structures [8].

Since engineering composite structures are commonly required to have natural frequencies away from the operating value to avoid structure resonance, stochastic natural frequency of composite structures has received increased attention in recent years. Oh and Librescu [9] adopted stochastic finite element method (SFEM) to investigate stochastic natural frequency of composite cantilever beams. Dey et al. [10,11] derived natural frequency statistics of composite plates by an RS-HDMR approach and later used a surrogate model. Chakraborty et al. [12] investigated stochastic free vibration behaviour of laminated composite plates using polynomial correlated function expansion and also response surface method (RSM). Tawfik et al. [13] presented a neural network-based second order reliability method to obtain probabilistic natural frequency of composite plates. Yin et al. [14] investigated stochastic variability of natural frequencies of laminated plates by modal stability procedure probabilistic approach. Singh et al. [15] provided a free vibration analysis of composite cylindrical panels with random material properties by first-order perturbation technique, and similar study is shown in [16] but for composite conical shell structure. It is noticed that these current studies are based on meso-scale (ply-scale) uncertainty such as uncertainty in ply elastic properties, ply orientation, ply density and thickness, and these meso-scale random variables are considered to be independent with one another. It is known that uncertainty of ply properties is essentially caused by stochastic variation of micro-scale (constituent-scale) parameters such as properties of fibre/matrix, fibre volume ratio, micro-structure, void, etc. [17]. Although probabilistic analysis based on meso-scale uncertainty takes less computation effort, uncertainty consideration started from micro-scale would provide insight into propagation of uncertainty across different scales [18] and also possible coupling between meso-scale random variables [19–24]. Still, it seems that little work has been reported about achieving comprehensive understanding on stochastic natural frequency of composite structures by considering uncertainty at different scales. Naskar et al. [25] showed difference in natural frequency statistics of composite plates derived from different scale uncertainty, but the uncertainties at different scales are somehow not connected. Therefore, it would be valuable to conduct comprehensive comparison on achieving natural frequency statistics of composite structures from meso-scale and micro-scale uncertainty, and any difference on the results would provide good indication or enlightenment to achieve complete description over uncertainty at different scales.

This study provides natural frequency analysis of typical composite structures based on micro-scale and meso-scale uncertainty. Meso-scale statistics are derived from micro-scale uncertainty by a combination of Monte-Carlo simulation and micromechanical model, and especially statistical correlation between meso-scale random variables is discussed. Response surface method (RSM) based on finite element modeling is employed to obtain approximate natural frequency of structures with complex shape or boundary condition. Statistics of natural frequencies of laminate plate and cylindrical shell are derived by a combination of Monte-Carlo simulation and the RSM. Statistics of structure natural frequency derived from different uncertainty considerations are compared and discussed.

## 2. Material Statistics

### 2.1. Micro-Scale Statistics

Statistics of mechanical and physical properties of common E-glass fibre and epoxy are shown in Table 1. Random variables considered in micro-scale mainly include $E_f$ (Young's modulus of fibre), $v_f$ (Poisson's ratio of fibre), $\rho_f$ (density of fibre), $E_m$ (Young's modulus of matrix), $v_m$ (Poisson's ratio of matrix), $\rho_m$ (density of matrix) and $V_f$ (fibre volume ratio). It has been experimentally shown that E-glass fibre tends to be mechanically isotropic [26]. Experimental observation on the statistics of fibre and matrix properties is generally rarely reported. In Table 1, the coefficient of variation (standard deviation/mean value) is assumed to be 5% if a reference is not given. All micro-scale properties shown in Table 1 are assumed to follow normal distribution in this study.

**Table 1.** Statistics of properties of fibre and matrix.

| | Mean Value | Standard Deviation | Distribution Type |
|---|---|---|---|
| **E-Glass Fibre** | | | |
| $E_f$ | 72 GPa | 5 GPa [27] | Normal |
| $v_f$ | 0.25 | 0.05 [26] | Normal |
| $\rho_f$ | 2540 kg/m$^3$ | 127 kg/m$^3$ | Normal |
| **Epoxy** | | | |
| $E_m$ | 3.45 GPa | 0.086 GPa [27] | Normal |
| $v_m$ | 0.35 | 0.0175 | Normal |
| $\rho_m$ | 1100 kg/m$^3$ | 55 kg/m$^3$ | Normal |
| **Fibre Volume Ratio** | | | |
| $V_f$ | 0.60 | 0.03 | Normal |

### 2.2. Meso-Scale Statistics

Statistics of meso-scale properties of unidirectional (UD) E-glass/epoxy are derived from micro-scale properties by a combination of Monte-Carlo simulation and bridging micromechanical model [28]. For E-glass/epoxy composite, the bridging micromechanical model is expressed as:

$$E_1 = E_f V_f + E_m V_m \tag{1}$$

$$E_{22} = \frac{(V_f + V_m a_{11})(V_f + V_m a_{22})}{(V_f + V_m a_{11})(V_f S_{f22} + a_{22} V_m S_{m22}) + V_f V_m (S_{m12} - S_{f12}) a_{12}} \tag{2}$$

$$\nu_{12} = \nu_f V_f + \nu_m V_m \tag{3}$$

$$G_{12} = \frac{(G_f + G_m) + V_f (G_f - G_m)}{(G_f + G_m) - V_f (G_f - G_m)} \tag{4}$$

where $a_{11}, a_{22}, a_{12}, S_{f11}, \ldots, S_{m12}$ are functions of the elastic properties of fibre and matrix as given in Huang [28]; $V_m$ denotes matrix volume ratio ($V_m = 1 - V_f$). $G_m(G_f)$ is the shear modulus of matrix (fibre) which is dependent on $E_m$ and $\nu_m$ ($E_f$ and $\nu_f$) due to isotropy of matrix (fibre). $E_1$ denotes normal modulus in the fibre direction; $E_2$ denotes the normal modulus in the transverse direction; $\nu_{12}$ denotes the in-plane major Poisson's ratio; $G_{12}$ denotes the in-plane shear modulus. In a composite material coordinate, 1 denotes fibre direction and 2 denotes transverse direction (perpendicular to fibre direction). Lamina density is derived by the rule of mixture which is written as:

$$\rho = \rho_f V_f + \rho_m V_m \tag{5}$$

In Monte-Carlo method, the accuracy ($\zeta$) on the estimation of a parameter $P$ is expressed as [29]:

$$\varsigma \approx \frac{s/\sqrt{N}}{\hat{P}} \tag{6}$$

where $\hat{P}$ is estimation of $P$, $N$ is sampling number, and $s$ is standard deviation of $P$ which is estimated as:

$$s = \sqrt{\frac{1}{N}\sum_{k=1}^{N} g(\mathbf{X}_k)^2 - \left(\frac{1}{N}\sum_{k=1}^{N} g(\mathbf{X}_k)\right)^2} \tag{7}$$

where $X_k$ is the $k$-th sample of random variables, and $g(X_k)$ is a function on $X_k$ such as Equations (1)–(5). $\zeta$ represents the relative difference between the mathematical expectation of a group of samples and the true mean value of corresponding random variable. Equation (6) indicates that a large enough number of samples would provide very accurate estimation on random parameters in Monte-Carlo simulation. In this work, $1 \times 10^5$ group of samples were drawn for the micro-scale random variables listed in Table 1 and they were substituted into Equations (1)–(5) to derive statistics of meso-scale properties. According to Equation (6), the sampling number of $1 \times 10^5$ provides estimation accuracy on meso-scale random variables better than 0.02%. The derived mean values, coefficient of variation (CoV) and linear correlation coefficients of UD E-glass/epoxy are shown in Table 2. The probabilistic distribution of derived ply mechanical properties was determined by Kolmogorov–Smirnov and Anderson–Darling tests, and it was shown that $E_{11}$, $E_{22}$, $\nu_{12}$, $G_{12}$ and $\rho$ all follow normal distribution with a confidence level at 95%.

**Table 2.** The statistics of UD E-glass/Epoxy composite.

| Random Variables | Linear Correlation Coefficient | | | | | Mean | CoV | Experiment [30] |
|---|---|---|---|---|---|---|---|---|
| | $E_1$ | $E_2$ | $G_{12}$ | $\upsilon_{12}$ | $\rho$ | | | |
| $E_1$ (GPa) | 1 | | | | | 44.5 | 0.083 | 45.6 |
| $E_2$ (GPa) | 0.65 | 1 | | | | 13.2 | 0.077 | 16.2 |
| $G_{12}$ (GPa) | 0.63 | 0.98 | 1 | | | 4.93 | 0.078 | 5.83 |
| $\upsilon_{12}$ | −0.06 | −0.10 | −0.20 | 1 | | 0.29 | 0.107 | 0.28 |
| $\rho$ (kg/m$^3$) | 0.43 | 0.72 | 0.71 | −0.07 | 1 | 1964 | 0.029 | |

Table 2 shows that significant statistical correlation exists between ply mechanical properties ($E_1$, $E_2$, $G_{12}$, $\upsilon_{12}$), which agrees with observation in [19,20,22]. A new observation is that statistical correlation between ply density ($\rho$) and ply elastic properties ($E_1$, $E_2$, $G_{12}$) was also significant, and this correlation seems to be rarely reported or noticed in previous studies. Actually, experimental data in Huang [28] provided a clear positive correlation between $V_f$ (fibre volume ratio) and $E_1$, $E_2$, $G_{12}$. Since it is known that ply density is positively dependent on the $V_f$, the theoretically predicted positive correlation between ply density and $E_1$, $E_2$, $G_{12}$ shown Table 2 is reasonable. Statistical correlation associated with ply density may affect stochastic vibration of composite structures as structure vibration is known to be mass dependent. Studies in later sections will demonstrate the effect of statistical correlation between $E_1$, $E_2$, $G_{12}$ and $\rho$ on the stochastic free vibration of composite structures.

## 3. Analysis of Stochastic Natural Frequency

### 3.1. Laminated Plate

In this work, stochastic natural frequency of several different configurations of laminated plates subjected to different boundary conditions as listed in Table 3 was investigated, including cross-ply laminate, angle-ply laminate, quasi-isotropic laminate and laminate of random ply orientations. For each laminate configuration, statistics of first order natural frequency were derived in three scenarios: (a) considering micro-scale uncertainty of constituent material properties as listed in Table 1, termed in short as 'micro-scale uncertainty'; (b) considering meso-scale uncertainty as listed in Table 2, and their statistical correlation was also considered, termed in short as 'correlated meso-scale uncertainty'; (c) considering meso-scale uncertainty of ply properties as listed in Table 2, but the statistical correlation between ply mechanical properties and also the density was neglected, termed in short as 'independent meso-scale uncertainty'. The flowchart of calculation process of stochastic natural frequency by considering the three different scenarios of uncertainty is shown in Figure 1. It is important to notice that most current studies (as introduced in Section 1) were based on 'independent meso-scale uncertainty', and it is necessary to discuss whether 'independent meso-scale uncertainty' provides reasonable estimation on the statistics of resonance frequency of composite structures.

**Table 3.** Laminate plate configurations.

| No. | Ply Thickness | Ply Orientation | Size | Boundary Condition |
|---|---|---|---|---|
| Laminate 1 | 0.45 mm | $(0/90)_4$ | 0.6 × 0.2 m | Simply supported |
| Laminate 2 | 0.45 mm | $[(\pm 60)_2]$s | 0.6 × 0.2 m | Simply supported |
| Laminate 3 | 0.45 mm | [0/±45/90]s | 0.6 × 0.2 m | Clamped |
| Laminate 4 | 0.45 mm | [15/45/60/90/30/0/10/20] | 0.6 × 0.2 m | Clamped |

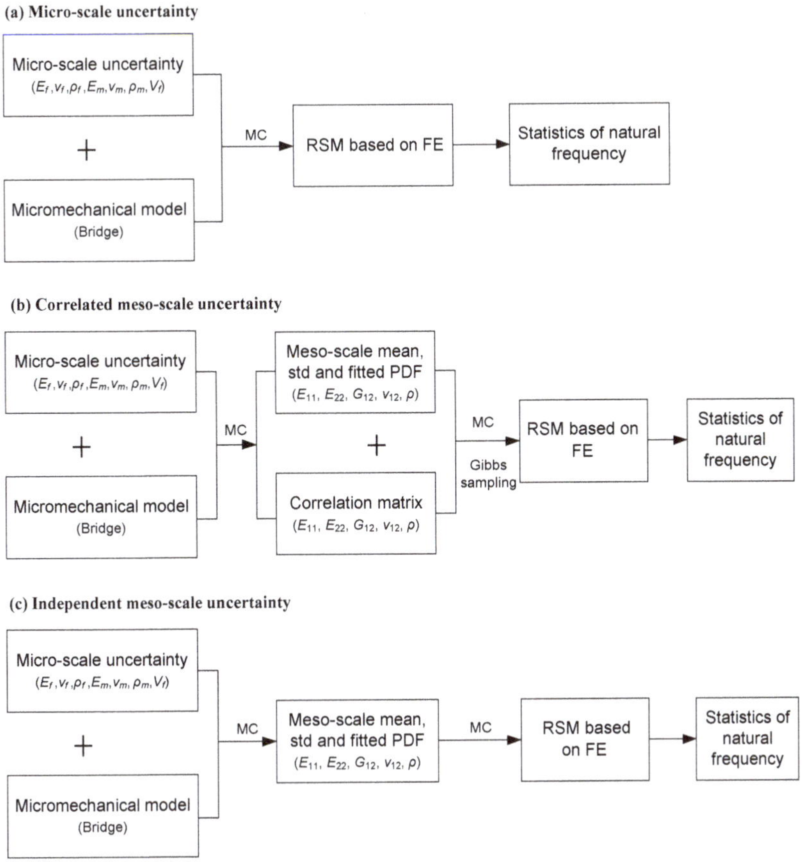


**Figure 1.** Flowchart of calculation of natural frequency statistics by different uncertainty consideration.

The statistics of first order natural frequency were derived by a combination of Monte-Carlo simulation and FE modeling. Four-node linear-elastic shell element (Shell181 in Ansys 14.5) is used to model the laminate. A mesh convergence study was conducted, which showed that a mesh size of $0.01 \times 0.01$ m provided a convergent solution on the first order natural frequency. Considering heavy computation cost of direct Monte-Carlo simulation on FE modeling, response surface method (RSM) with quadratic expression was employed to derive an approximate surrogate model to map random variables to the first order natural frequency, as shown in Equation (8):

$$g(\mathbf{X}) = a + b_i \sum_{i=1}^{n} x_i + c_{ij} \sum_{i=1}^{n} x_i \sum_{i=1}^{n} x_j \tag{8}$$

where $g$ represents the first order natural frequency, $x_i$ represents micro-scale or meso-scale random variable shown in Table 1 or Table 2, and $a$, $b$ and $c$ are regression terms. In this work, the regression terms were determined by central composite design method [31]. The central composite design was composed of one central point, two axis points per input variable and factorial points at the corners of the hypercube. Twenty-seven design points were employed for the meso-scale uncertainty scenario and 79 design points were employed for the micro-scale uncertainty scenario. In comparison to other experimental design methods such as the Box–Behnken method, the central composite design method could provide high quality predictions over the entire design space, and it is more flexible considering the number of input random variables [32].

For all laminate configurations shown in Table 3, $1 \times 10^5$ group samples of random variables were drawn to conduct stochastic analysis based on the surrogate model derived by the RSM, and corresponding relative error on estimation of the mean value of first order natural frequency was less than 0.01% (calculated by Equation (6)). For the scenario of 'correlated meso-scale uncertainty', Gibbs sampling was employed to draw samples of meso-scale random variables to account for their correlation [21], by the procedure as follows:

(1) Initiation of random variable $x_i$ ($i = 1, 2, \ldots, n$);
(2) For $t = 0, 1, 2, \ldots$, do the iterative sampling as follows:

1. $x_1^{(t+1)} \sim p(x_1|x_2^t, x_3^t, \cdots x_n^t)$
2. $x_2^{t+1} \sim p(x_2|x_1^{t+1}, x_3^t, \cdots x_n^t)$
3. .........
4. $x_j^{t+1} \sim p(x_j|x_1^{t+1}, \cdots, x_{j-1}^{t+1}, x_{j+1}^t \cdots x_n^t)$
5. .........
6. $x_n^{t+1} \sim p(x_n|x_1^{t+1}, x_2^{t+1}, \cdots x_{n-1}^{t+1})$

where $t$ denotes the number of sample, and $p()$ denotes probability density function.

Table 4 lists mean value and standard deviation of derived first order natural frequency of different laminate plates. It is seen that mean values of the first order natural frequency derived from the three different consideration of uncertainty were almost identical, with a difference less than 0.01%. The standard deviations of the first order natural frequency derived from 'micro-scale uncertainty' and 'correlated meso-scale uncertainty' were also very similar, with a difference less than 2%. This small difference could be introduced by the process of probabilistic density function fitting of meso-scale random variables. However, significant difference was seen between the standard deviation derived by 'independent meso-scale uncertainty' and 'correlated meso-scale uncertainty' or 'micro-scale uncertainty', where the standard deviation derived by 'independent meso-scale uncertainty' was around 10% larger.

**Table 4.** Statistics of first order natural frequencies of laminate plate (unit in Hz).

| No. | Micro-Scale | | Correlated Meso-Scale | | Independent Meso-Scale | |
|---|---|---|---|---|---|---|
| | Mean | Std | Mean | Std | Mean | Std |
| Laminate 1 | 167.2 | 5.17 | 167.3 | 5.19 | 167.3 | 5.61<br>8.1% * |
| Laminate 2 | 181.0 | 5.89 | 181.0 | 6.00 | 181.0 | 6.53<br>8.8% * |
| Laminate 3 | 290.5 | 8.12 | 290.5 | 8.25 | 290.7 | 8.72<br>5.7% * |
| Laminate 4 | 275.6 | 7.65 | 275.7 | 7.88 | 275.8 | 8.88<br>12.7% * |

* Difference from its counterpart value of 'Correlated Meso-Scale'.

Figure 2 shows histogram and probability distribution of the first order natural frequency of Laminate 1 derived from 'independent meso-scale uncertainty' and 'correlated meso-scale uncertainty'. From the histogram shown in Figure 2a, it is shown that the 'correlated meso-scale uncertainty' provides more samples at the middle region and less samples at the tail region. By Kolmogorov–Smirnov and Anderson–Darling tests, the probability distribution of the first order natural frequency tended to follow normal distribution. From the fitted probability density function by normal distribution as shown in Figure 2b, it is seen that there is notable difference in the tail region between the two probability density functions. The cumulative probability of the first order natural frequency is shown in Figure 3, which shows that the 'independent meso-scale uncertainty' provides a notable overestimation on the cumulative probability. This demonstrates a necessity to account for statistical correlation between meso-scale uncertainty to derive accurate stochastic natural frequency.

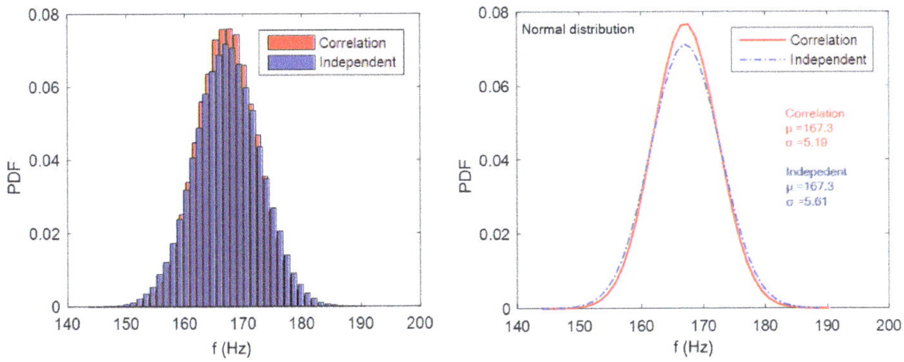


(a) (b)

**Figure 2.** Probabilistic distribution of first order frequency of Laminate 1 by considering correlated and independent meso-scale random variables: (**a**) histogram; (**b**) fitted by normal distribution.

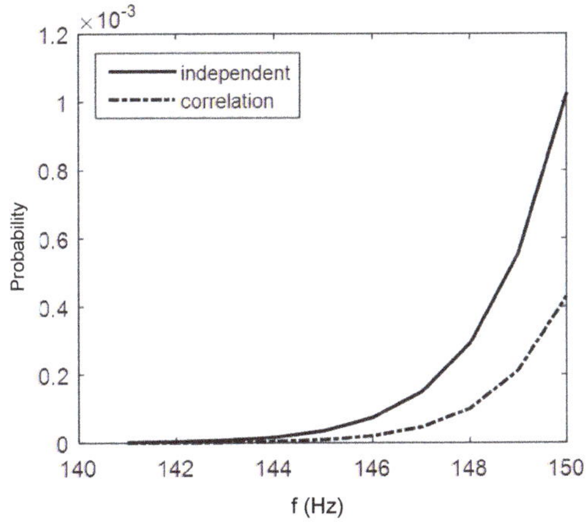


**Figure 3.** Cumulative probability of first order frequency of Laminate 1, derived by considering correlated and independent meso-scale random variables.

To achieve better understanding on the contribution of different uncertainty to the stochastic variation of the natural frequency, linear correlation coefficients between the first order natural frequency of Laminate 1 and meso-scale random variables were derived, as shown in Figure 4. The linear correlation coefficient reflected a stochastic dependence or sensitivity of stochastic variation of the natural frequency over input random variables. It is seen that neglecting correlation between meso-scale uncertainty suppressed the stochastic dependence on $E_{22}$ and $G_{12}$, which was similar to observation in [1] where stochastic deflection of a laminate structure was investigated. An interesting observation is on the dependence over laminate density. If the statistical correlation between the meso-scale input random variables was ignored, the stochastic dependence of the first order natural frequency over ply density was negative. It seems to be under expectation as it is known that larger density results into smaller natural frequency if plate stiffness and boundary conditions remain unchanged. However, from a multi-scale viewpoint, the ply density is actually positively correlated with ply modulus, and a larger ply density also indicates a larger stiffness. Hence, for the case of 'correlated meso-scale uncertainty', a positive correlation was seen between the natural frequency and the ply density. This highlights the importance to consider statistical correlation between meso-scale mechanical properties to achieve comprehensive understanding on the stochastic free vibration of laminate plates.

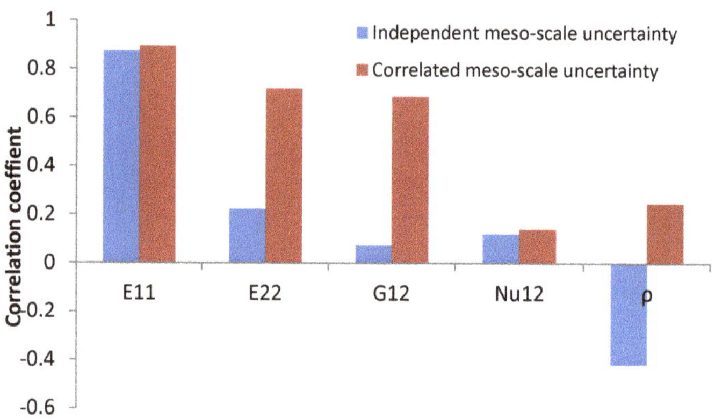


**Figure 4.** Linear correlation coefficient between first order natural frequency of laminate plate and meso-scale random variables.

For Laminates 2, 3 and 4 listed in Table 3, similar characteristics of the standard deviation, cumulative probability and sensitivity of the first order natural frequency were obtained as compared to Laminate 1, considering the discrepancy between the three different uncertainty considerations.

### 3.2. *Laminated Cylindrical Shell*

Laminated cylindrical shell structures are widely used in airspace, aerospace and energy engineering fields, with applications such as the dual launch system (SYLDA) and interstage skirt structure (ISS) of Ariane 5 launcher [33]. Especially, circular cut-outs are made on the laminated cylindrical shell of the SYLDA. In this section, stochastic natural frequency of a laminated cylindrical shell with circular cut-outs is investigated. The geometry of the laminated cylindrical shell is shown in Figure 5a, which was based on a scaled model of the SYLDA of Ariane 5 launcher where the shell diameter is equal to the shell length and the cut-out diameter is around 0.1–0.2 times of shell length. The laminate was set as eight-layered E-glass fibre/epoxy material, and the ply orientation was $(\pm\theta)_4$. A ply with an orientation at 0 indicates that fibre is aligned in the cylinder length direction. Ply thickness was selected as 0.45 mm. Both ends of the laminated cylindrical shell were clamped. FE modeling was employed to obtain its first order natural frequency, as shown in Figure 5. The laminate was modeled by four-node linear-elastic shell element (Shell181 in Ansys 14.5), and a mesh size of 5 × 5 mm was found to be fine enough to provide a convergent solution. FE solution on the first order vibration mode of $(\pm 45)_4$ laminate cylindrical shell is shown in Figure 5b.

Similar to the analysis of laminate plates in last section, a combination of RSM and Monte-Carlo simulation is adopted to obtain the statistics of the first order natural frequency, by considering 'micro-scale uncertainty', 'independent meso-scale uncertainty' and 'correlated meso-scale uncertainty'. The derived standard deviation of the first order natural frequency is shown in Figure 6. It is seen that the difference on the standard deviation between the three difference uncertainty consideration was fairly small for θ at 0°, 30°, 45° and 60°. However, for the situation of θ = 90°, the standard deviation derived from 'independent meso-scale uncertainty' was 10.4% larger than that of 'correlated meso-scale uncertainty' or 'micro-scale uncertainty'. This difference agrees with that of laminate plates shown in Table 4, which again demonstrates that ignoring the statistical correlation between meso-scale uncertainty probably leads to overestimation of the standard deviation of structure natural frequency.

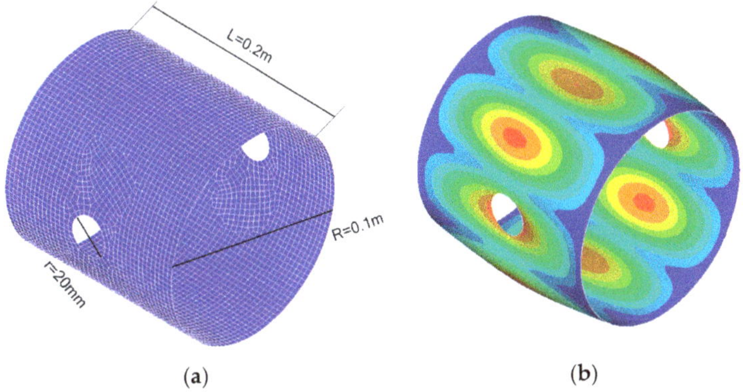


**Figure 5.** (**a**) Geometrical dimension and mesh of laminate cylindrical shell; (**b**) first order vibration model of $(\pm 45)_4$ laminate.

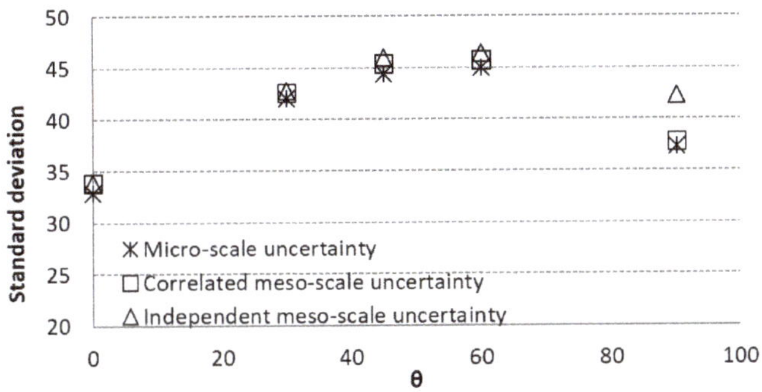


**Figure 6.** Standard deviation of first order natural frequency of $(\pm\theta)_4$ laminate cylindrical shell derived by considering different uncertainty.

## 4. Discussion

Fibre reinforced composite is essentially a type of multi-scale material. Uncertainty consideration from different scales would provide more comprehensive understanding on accurate uncertainty characterization. Previous work has highlighted possible existence of significant correlation between ply mechanical properties [19,22]. This correlation would affect reliability of composite structures considering structure failure defined by structure deformation, fracture or buckling [20]. It is known that natural frequency of composite structures not only depends on ply elastic properties but also laminate density. Therefore, as a new observation on the statistical correlation associated to ply density, it would be interesting to see how much the correlation contributes to the statistics of structure natural frequency. Here, a further comparison is conducted on deriving the statistics of structure natural frequency by considering two scenarios: (1) considering only correlation between ply mechanical properties ($E_{11}$, $E_{22}$, $G_{12}$, $\nu_{12}$); (2) considering correlation between ply mechanical properties and also density ($E_{11}$, $E_{22}$, $G_{12}$, $\nu_{12}$, $\rho$). The derived standard deviations of the first order natural frequency of Laminate 1 (see Table 3) and $(90)_8$ laminate cylindrical shell are shown in Table 5. It is interesting to observe that neglecting correlation associated with ply density results into 30–50% overestimation of the standard deviation, while neglecting all correlation between meso-scale uncertainty results into around 10% overestimation.

**Table 5.** Standard deviation of first order natural frequency considering different uncertainty.

| | Meso-Scale Correlation ($E_{11}$, $E_{22}$, $G_{12}$, $\nu_{12}$, $\rho$) | Meso-Scale Correlation ($E_{11}$, $E_{22}$, $G_{12}$, $\nu_{12}$) | Meso-Scale Independent |
|---|---|---|---|
| $(0/90)_8$ plate | 5.10 Hz<br>– | 6.70 Hz<br>(31.4%) * | 5.61 Hz<br>(10.0%) * |
| $(90)_8$ cylindrical shell | 37.8 Hz<br>– | 56.0 Hz<br>(48.2%) * | 42.2 Hz<br>(11.6%) * |

* Difference from its counterpart value of 'meso-scale correlation ($E_{11}$, $E_{22}$, $G_{12}$, $\nu_{12}$, $\rho$)'.

Overestimation of standard deviation by 30–50% would lead to very large error in corresponding cumulative probability estimation. Therefore, it is important to account for the statistical correlation associated with ply density to obtain accurate stochastic free vibration behaviour of laminated structures.

A recent study [25] reports that larger standard deviation of natural frequency was obtained for the case of 'micro-scale uncertainty' than that of the 'independent meso-scale uncertainty', which seems to contradict with our observations in Table 4 and Figure 5. Actually, the comparison in [25] was conducted by assuming identical coefficient of variation (CoV) for both meso-scale and micro-scale uncertainties, for example the CoV of all of meso-scale and micro-scale random variables were assumed to be 5%. However, Tables 1 and 2 tell that micro-scale uncertainty with CoV at 5% actually results in CoV of meso-scale uncertainty at around 10%. The uncertainty amplification from micro-scale to meso-scale was also addressed in [18]. The uncertainty amplification explains why larger standard deviation was obtained for the case of 'micro-scale uncertainty' in [25]. Meanwhile, statistical correlation between meso-scale uncertainty was not considered and discussed in [25].

## 5. Conclusions

The present study provides statistical natural frequency of composites structures derived from micro-scale and meso-scale uncertainty, for the circumstance composite structures are composed of plies with unidirectional fibre reinforcement. Meso-scale uncertainty is obtained from micro-scale uncertainty by a combination of Monte-Carlo simulation and bridging micromechanical model. It is observed that significant statistical correlation may not only exist between ply elastic properties but also associated with ply density. The statistical correlation between meso-scale physical properties is commonly ignored in current study on stochastic free vibration of composite structures. Comparison is conducted on natural frequency statistics of composite plates and cylindrical shell considering 'micro-scale uncertainty', 'correlated meso-scale uncertainty' and 'independent meso-scale uncertainty'. Several conclusions are drawn as follows:

1. Natural frequency statistics of composite structures could be reasonably derived from either micro-scale random variables or meso-scale random variables but their statistical correlation needs to be well characterised;
2. Ignoring the statistical correlation between meso-scale uncertainty may result in large overestimation of standard deviation of structure natural frequency;
3. If the natural frequency statistics of composite structures are calculated by meso-scale uncertainty, it is especially important to consider the statistical correlation associated with ply density, which is somehow not noticed by previous studies. It also shows that dependence of the stochastic variation of natural frequency over ply density is positive rather than negative.

It is very important to achieve accurate statistics of structure natural frequency in engineering design, so that potential structure resonance could be well avoided. Therefore, this work further suggests establishing standard experimental approaches to obtain statistical correlation between ply elastic properties and ply density.

**Author Contributions:** Conceptualization, S.Z. and X.C.; methodology, S.Z.; software, S.Z.; formal analysis, S.Z.; investigation, S.Z. and X.C.; writing—original draft preparation, S.Z.; writing—review and editing, X.C.; funding acquisition, S.Z.

**Funding:** This research was funded by National Natural Science Foundation of China, grant number 51405501.

**Conflicts of Interest:** The authors declare no conflict of interest.

## References

1. Mustafa, G.; Suleman, A.; Crawford, C. Probabilistic micromechanical analysis of composite material stiffness properties for a wind turbine blade. *Compos. Struct.* **2015**, *131*, 905–916. [CrossRef]
2. Hwang, T.-K.; Hong, C.-S.; Kim, C.-G. Probabilistic deformation and strength prediction for a filament wound pressure vessel. *Compos. Part B Eng.* **2003**, *34*, 481–497. [CrossRef]
3. Chen, N.Z.; Sun, H.H.; Soares, C.G. Reliability analysis of a ship hull in composite material. *Compos. Struct.* **2003**, *62*, 59–66. [CrossRef]
4. Frangopol, D.M.; Recek, S. Reliability of fiber-reinforced composite laminate plates. *Probab. Eng. Eng. Mech.* **2003**, *18*, 119–137. [CrossRef]
5. Chamis, C.C.; Abumeri, G.H. Probabilistic dynamic buckling of composite shell structures. *Compos. Part A Appl. Sci. Manuf.* **2005**, *36*, 1368–1380. [CrossRef]
6. Ganesan, R.; Kowda, V.K. Buckling of composite beam-columns with stochastic properties. *J. Reinf. Plast. Compos.* **2005**, *24*, 513–543. [CrossRef]
7. Fan, Z.; Jiang, Y.; Zhang, S.; Chen, X. Experimental research on vibration fatigue of cfrp and its influence factors based on vibration testing. *Shock Vib.* **2017**, *2017*, 18. [CrossRef]
8. Chiachio, M.; Chiachio, J.; Rus, G. Reliability in composites—A selective review and survey of current development. *Compos. Part B Eng.* **2012**, *43*, 902–913. [CrossRef]
9. Oh, D.H.; Librescu, L. Free vibration and reliability of composite cantilevers featuring uncertain properties. *Reliab. Eng. Syst. Safe.* **1997**, *56*, 265–272. [CrossRef]
10. Dey, S.; Mukhopadhyay, T.; Adhikari, S. Stochastic free vibration analysis of angle-ply composite plates—A rs-hdmr approach. *Compos. Struct.* **2015**, *122*, 526–536. [CrossRef]
11. Dey, S.; Naskar, S.; Mukhopadhyay, T.; Gohs, U.; Spickenheuer, A.; Bittrich, L.; Sriramula, S.; Adhikari, S.; Heinrich, G. Uncertain natural frequency analysis of composite plates including effect of noise—A polynomial neural network approach. *Compos. Struct.* **2016**, *143*, 130–142. [CrossRef]
12. Chakraborty, S.; Mandal, B.; Chowdhury, R.; Chakrabarti, A. Stochastic free vibration analysis of laminated composite plates using polynomial correlated function expansion. *Compos. Struct.* **2016**, *135*, 236–249. [CrossRef]
13. Tawfik, M.E.; Bishay, P.L.; Sadek, E.A. Neural network-based second order reliability method (nnbsorm) for laminated composite plates in free vibration. *Comp. Model. Eng. Sci.* **2018**, *115*, 105–129.
14. Yin, Q.; Lardeur, P.; Druesne, F. Performances assessment of the modal stability procedure for the probabilistic free vibration analysis of laminated composite structures. *Compos. Struct.* **2018**, *203*, 474–485. [CrossRef]
15. Singh, B.N.; Yadav, D.; Iyengar, N.G.R. Free vibration of composite cylindrical panels with random material properties. *Compos. Struct.* **2002**, *58*, 435–442. [CrossRef]
16. Tripathi, V.; Singh, B.N.; Shukla, K.K. Free vibration of laminated composite conical shells with random material properties. *Compos. Struct.* **2007**, *81*, 96–104. [CrossRef]
17. Sriramula, S.; Chryssanthopoulos, M.K. Quantification of uncertainty modelling in stochastic analysis of frp composites. *Compos. Part A Appl. Sci. Manuf.* **2009**, *40*, 1673–1684. [CrossRef]
18. Rollet, Y.; Bonnet, M.; Carrère, N.; Leroy, F.H.; Maire, J.F. Improving the reliability of material databases using multiscale approaches. *Compos. Sci. Technol.* **2009**, *69*, 73–80. [CrossRef]
19. Shaw, A.; Sriramula, S.; Gosling, P.D.; Chryssanthopoulos, M.K. A critical reliability evaluation of fibre reinforced composite materials based on probabilistic micro and macro-mechanical analysis. *Compos. Part B Eng.* **2010**, *41*, 446–453. [CrossRef]

20. Zhang, S.; Zhang, C.; Chen, X. Effect of statistical correlation between ply mechanical properties on reliability of fibre reinforced plastic composite structures. *J. Compos. Mater.* **2015**, *49*, 2935–2945. [CrossRef]
21. Zhang, S.; Zhang, L.; Wang, Y.; Tao, J.; Chen, X. Effect of ply level thickness uncertainty on reliability of laminated composite panels. *J. Reinf. Plast. Compos.* **2016**, *35*, 1387–1400. [CrossRef]
22. Zhang, S.; Wang, H.; Zhang, L.; Chen, X. Statistical correlation between elastic properties of plain-weave composite and its influence on structure reliability. *Compos. Struct.* **2018**, *200*, 939–945. [CrossRef]
23. Zhou, X.Y.; Gosling, P.D.; Ullah, Z.; Kaczmarczyk, Ł.; Pearce, C.J. Exploiting the benefits of multi-scale analysis in reliability analysis for composite structures. *Compos. Struct.* **2016**, *155*, 197–212. [CrossRef]
24. Zhou, X.Y.; Gosling, P.D.; Ullah, Z.; Kaczmarczyk, L.; Pearce, C.J. Stochastic multi-scale finite element based reliability analysis for laminated composite structures. *Appl. Math. Model.* **2017**, *45*, 457–473. [CrossRef]
25. Naskar, S.; Mukhopadhyay, T.; Sriramula, S.; Adhikari, S. Stochastic natural frequency analysis of damaged thin-walled laminated composite beams with uncertainty in micromechanical properties. *Compos. Struct.* **2017**, *160*, 312–334. [CrossRef]
26. Mounier, D.; Poilâne, C.; Bûcher, C.; Picart, P. Evaluation of transverse elastic properties of fibers used in composite materials by laser resonant ultrasound spectroscopy. In Proceedings of the 11th Congrès Français d'Acoustique (Acoustics 2012), Nantes, France, 23–27 April 2012; Available online: https://hal.archives-ouvertes.fr/hal-00811303/document (accessed on 6 June 2019).
27. Sanei, S.H.R.; Fertig Iii, R.S. Length-scale dependence of variability in epoxy modulus extracted from composite prepreg. *Polym. Test* **2016**, *50*, 297–300. [CrossRef]
28. Huang, Z. Micromechanical prediction of ultimate strength of transversely isotropic fibrous composites. *Int. J. Solids Struct.* **2001**, *38*, 4147–4172. [CrossRef]
29. Rubinstein, R.Y.; Kroese, D.P. *Simulation and the Monte Carlo Method*, 2nd ed.; John Wiley & Sons: Hoboken, NJ, USA, 2007.
30. Soden, P.D.; Hinton, M.J.; Kaddour, A.S. Lamina properties, lay-up configurations and loading conditions for a range of fibre-reinforced composite laminates. *Compos. Sci. Technol.* **1998**, *58*, 1011–1022. [CrossRef]
31. Anderson, M.J.; Whitcomb, P.J. *RSM simplified: Optimizing Processes Using Response Surface Methods for Design of Experiments*, 2nd ed.; Taylor & Francis Group: London, UK, 2016.
32. Reh, S.; Beley, J.-D.; Mukherjee, S.; Khor, E.H. Probabilistic finite element analysis using ansys. *Struct. Saf.* **2006**, *28*, 17–43. [CrossRef]
33. Alfano, M.; Bisagni, C. Probability-based methodology for buckling investigation of sandwich composite shells with and without cut-outs. *Int. J. Comput. Meth. Eng. Sci. Mech.* **2017**, *18*, 77–90. [CrossRef]

applied sciences

MDPI

Article

# Seismic Response of Steel Moment Frames (SMFs) Considering Simultaneous Excitations of Vertical and Horizontal Components, Including Fling-Step Ground Motions

**Shahrokh Shahbazi [1], Armin Karami [2], Jong Wan Hu [3,4,*] and Iman Mansouri [5]**

1 TAAT Investment Group, Tehran 18717-13553, Iran; shahrokh.shahbazi25@yahoo.com
2 Department of Civil, Environmental and Land Management Engineering, Politecnico di Milano, Milan 20133, Italy; armin.karami@mail.polimi.it
3 Department of Civil and Environmental Engineering, Incheon National University, Incheon 22012, Korea
4 Incheon Disaster Prevention Research Center, Incheon National University, Incheon 22012, Korea
5 Department of Civil Engineering, Birjand University of Technology, Birjand 97175-569, Iran; mansouri@birjandut.ac.ir
* Correspondence: jongp24@incheon.ac.kr; Tel.: +82-32-835-8463

Received: 16 March 2019; Accepted: 14 May 2019; Published: 20 May 2019

**Abstract:** Near-field (NF) earthquakes have drawn considerable attention from earthquake and structural engineers. In the field of earthquake engineering, numerous studies have identified the devastating nature of such earthquakes, and examined the characteristics related to the response of engineering structures to these types of earthquakes. Herein, special steel moment frames (SMFs) of three-, five-, and eight-story buildings have been examined via a nonlinear time history analysis in OpenSees software. The behavioral seismic differences of these frames have been evaluated in two states: (1) under the simultaneous excitation of the horizontal and vertical constituents of near-field earthquakes that have Fling-steps in their records; and (2) under simultaneous excitation of the horizontal and vertical constituents of far-field (FF) earthquakes. In addition, during modeling, the effects of panel zones have been considered. Considering that the simultaneous effects of the horizontal and vertical constituents of near-field earthquakes were subjected to a fling-step resulting in an increased inter-story drift ratio, the horizontal displacement of stories, an axial force of columns, created the moment in columns, base shearing of the structure, and velocity and acceleration of the stories.

**Keywords:** near-field earthquake; fling-step; far-field; simultaneous excitation; special moment frame (SMF)

## 1. Introduction

According to the elastic rebound theory, the sudden release of gradually accumulated stress and displacement in the earth's crust leads to the formation of a large permanent ground displacement over a few seconds and creates a fling-step [1] (Figure 1). The result would be a long-period pulse which is different from the better-known directivity pulse produced by the constructive interference of propagating seismic waves. Ventura et al. [2], and also Grimaz and Malisan [3], investigated the influences of fling-step on buildings through a parametric study; their study proves that the motion with the fling effect creates a much larger response than those without the fling effect (and thus potentially greater damage). Moreover, several studies have shown a considerable difference between the seismic demands of near- and far-field records. This is due to the high frequency content of near-field records and pulse-like behavior of these types of ground motions [4–8].

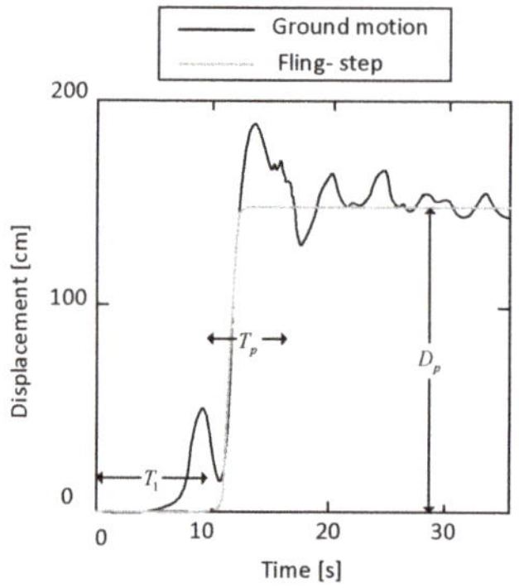


**Figure 1.** Ground motion depicting fling-step (i.e., static offset) from the YPT-N/S station in the 1999 Kocaeli, Turkey earthquake. $D_p$: Displacement amplitude; $T_p$: Period or duration; $T_1$: arrival time.

The main purpose of this paper is to develop a study of reference [8]. Kalkan and Kunnath [8] considered only the horizontal component of ground motion, while in this paper the simultaneous excitations of the horizontal and vertical constituents are considered. Besides, the panel zones are modeled in this paper, which were ignored in reference [8]. Among the seismic demands, only inter-story drift has been selected in the study of Kalkan and Kunnath, however in the present paper, in addition to inter-story drift, story displacement, columns' axial force, columns' moment, story shear, velocity and acceleration of story through an extensive post-processing are investigated.

Numerous studies have sought to clarify the response of buildings to diverse kinds of abbreviated pulses and create a design method for pulse-type records. For instance, references [9–11] evaluated the response of buildings and predicted the near-field response spectrum. References [12–17] examined the effects of pulse-like motions on building response. Moreover, the impacts on strength reduction factors were assessed in references [9,15,18,19]. Recently, Shahbazi et al. [14] studied the effects of soil classification on the seismic behavior of steel moment frames (SMFs), considering soil structure interactions and near-field ground motions.

An exclusive sine wave model for fling-step pulse waveforms was recently introduced by Burks and Baker [20]. However, a single sine wave cannot fully describe fling-step pulses, as they may consist of more than one frequency. Burks and Baker [20] presented a number of equations to predict the parameters of their model without the need for producing fling-step pulses.

Since these techniques take a lot of time, Farid Ghahari et al. [21] extracted the pulse component of a ground motion by adopting the moving average method. Soil structure interactions (SSIs) can also affect the response of buildings. The seismic evaluation of SMFs subjected to near-field ground motions with a forward directivity, including SSI impacts, were investigated by Shahbazi et al. [22], and the effect of fling-steps were ignored.

Moment frame systems are simply defined as assemblages of rigidly connected columns and beams [23–26], with its lateral strength mainly being achieved by the creation of bending moments and shear forces in the members and joints [27–30]. Owing to their pulse-like properties, near-field earthquakes result in a fast transmission of energy from the ground to the structure. Therefore, the elements in charge of increasing structural ductility fail to optimally perform. Thus, this study selected special steel moment frames, which have high ductility expectations. Moreover, in order to consider more realistic conditions in the modeling process, not only were the horizontal and vertical effects of near- and far-fault records simultaneously applied to the frames (which is the case in real earthquakes), but the effects of panel zone modeling were also incorporated.

## 2. The Effects and Characteristics of Near-Field Earthquakes

Past studies show that near-field earthquakes have the critical energy of pulses. However, while these earthquakes may have a smaller magnitude than Richter, they have a high potential for damage. In general, near-field earthquakes have specific characteristics, which are briefly mentioned as follows:

(1) The velocity and acceleration history of these earthquakes have long-period pulses.
(2) The ratio of maximum velocity to the maximum acceleration of ground is significant in these earthquakes.
(3) Sometimes near-field quakes lead to major permanent transformations of the earth [31].

These characteristics arise from the many phenomena that occur near the seismic zone, which will be discussed later.

### *Fling-Step Effect*

One of the recorded characteristics of the near-field records in last two decades, such as the 1999 earthquakes in Turkey and Taiwan, is the fling-step. This property occurred in the range of a few seconds of slip in the direction of the fault slip and is independent of the dynamic displacement originating from the directivity pulse of the ground's rupture. Accordingly, this phenomenon will affect components that are in the line of strike (parallel to the slip fault in strike-slip earthquakes, and in the direction of the slope in the dip-slip earthquakes).

For dip-slip earthquakes, fling-step and directivity effects occur in one direction and fling-step ground simultaneously occurs with maximum dynamic displacement, which must be considered as loads perpendicular to each other. The impact of the loads at one time is associated with a high potential for damage [32]. Briefly, fling-step ground is associated with a long-range velocity pulse and a uniform step in the time history of displacement. For instance, the Yarimka station experienced an approximately 2-meter permanent displacement during the Kjaili earthquake [33] (see Figure 2).

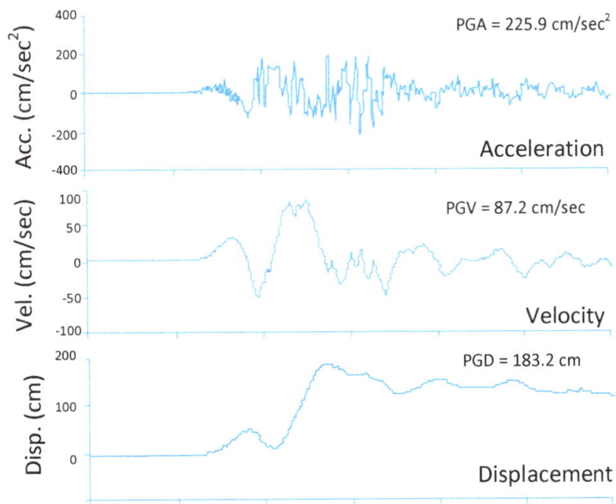


**Figure 2.** NS direction Yarimca station recordings during the Kocaeli earthquake [33].

The below figure is schematic of a strike-slip fault, the pulse is caused by the forward orientation and subsequent deformation which occurs in the perpendicular parameter and parallel to the major parameter of the fault, respectively. However, these two impacts are considered in the perpendicular parameters to the fault for dip-slip faults (see Figure 3), and time intervals where these factors are demonstrated with each other and separately are represented in Figure 4. Figure 3 displays the trend of pulse movement along with remnant displacement for strike-slip and dip-slip faults. The time interval plots caused by strike and reverse faults can be observed in Figure 4.

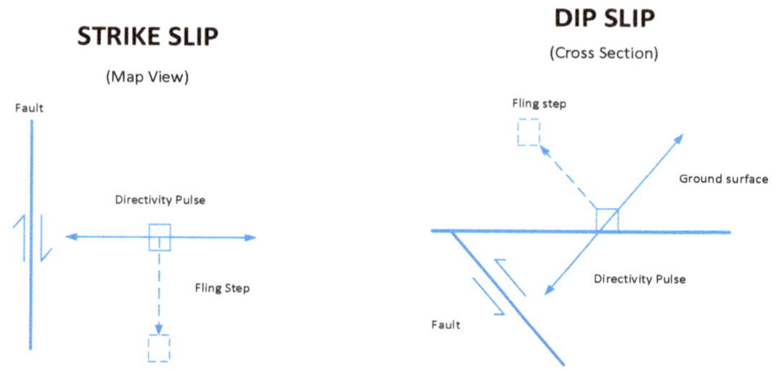


**Figure 3.** Conceptual plot representing the trends of the fling-step and directivity pulse for strike-slip and dip-slip faulting [32].

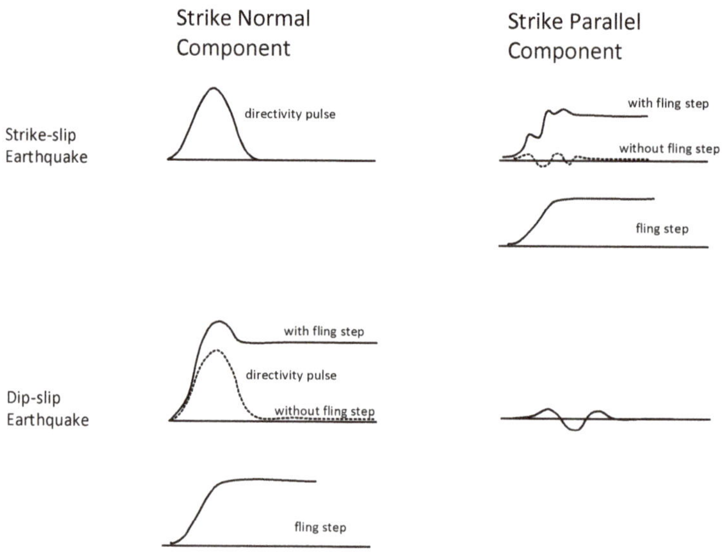


**Figure 4.** Conceptual plot showing the time intervals of strike-slip and dip-slip faulting in which the fling-step and directivity pulse are displayed together and discretely [32].

## 3. Numerical Models

### *3.1. Selected Numerical Models*

In the current research, steel moment skeletons of 3-, 5- and 8-floor with a particular ductility were applied for investigating the seismic behavior of three models founded on soil type 2 (375 m/s ≤ $V_s$ (shear-wave velocity) ≤ 750 m/s). Seismic design of these three models was conducted using ETABS software and Iranian national construction codes [34]. The height of stories is 3.2 m and each of them was founded with three linear and lateral spans by the same width of 5 m [35]. In addition, corresponding columns and beams were chosen on both sides with stories undergoing similar efficiencies. Based on European standard cross sections, various kind of profiles were assumed for beams and columns, yielding cross sections which were employed in this research for beams and box-shapes were suggested for columns. As a result, the schematic model of the reference skeletons is illustrated in Figure 5. The profiles of the beams and columns for three-, five- and eight-story skeletons are presented in Table 1. Furthermore, leaning columns were applied to modeling second order impacts (Figure 5).

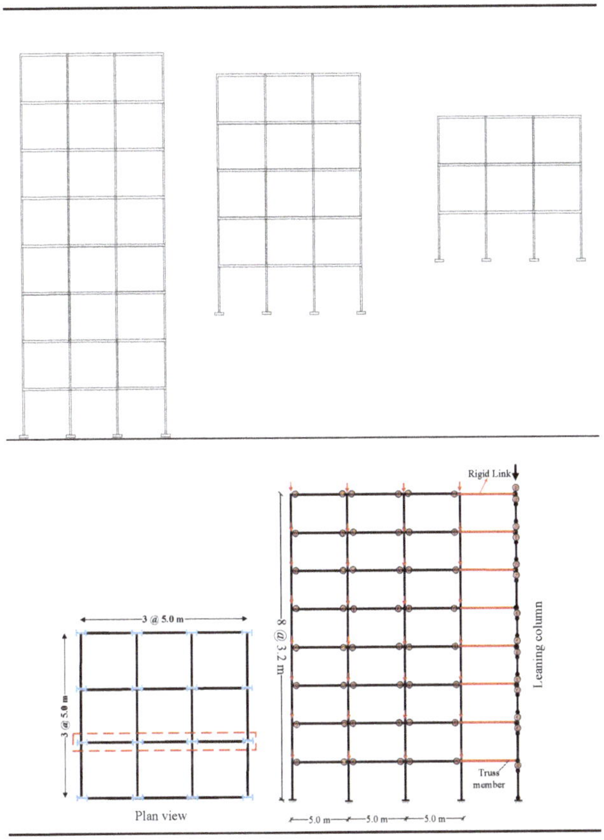


**Figure 5.** Topology and grouping details of stories [26].

**Table 1.** The section of structures.

| | No. | Column | Beam | | No. | Column | Beam |
|---|---|---|---|---|---|---|---|
| 3-story | 1 | Box 200 × 200 × 20 | IPE 300 | 8-story | 1 | Box 280 × 280 × 20 | IPE 400 |
| | 2 | Box 200 × 200 × 20 | IPE 300 | | 2 | Box 280 × 280 × 20 | IPE 400 |
| | 3 | Box 200 × 200 × 20 | IPE 270 | | 3 | Box 240 × 240 × 20 | IPE 360 |
| | 4 | Box 280 × 280 × 20 | IPE 400 | | 4 | Box 180 × 180 × 20 | IPE 330 |
| | 5 | Box 280 × 280 × 20 | IPE 300 | | 5 | Box 400 × 400 × 20 | IPE 400 |
| | 6 | Box 280 × 280 × 20 | IPE 270 | | 6 | Box 340 × 340 × 20 | IPE 450 |
| 5-story | 1 | Box 240 × 240 × 20 | IPE 330 | | 7 | Box 300 × 300 × 20 | IPE 400 |
| | 2 | Box 240 × 240 × 20 | IPE 360 | | 8 | Box 240 × 240 × 20 | IPE 360 |
| | 3 | Box 180 × 180 × 20 | IPE 240 | | | | |
| | 4 | Box 300 × 300 × 20 | IPE 330 | | | | |
| | 5 | Box 300 × 300 × 20 | IPE 360 | | | | |
| | 6 | Box 240 × 240 × 20 | IPE 240 | | | | |

The following symbol is applied for the columns place in the analysis result section (in Section 4 of paper):
Cij* is the code for position of results in the columns;
i = story number;
j = number of columns located in the left-hand of buildings.

There are different assumptions being considered in this research. Dead (fixed) load on a building is due to the weight of constant elements including beams, floor segments, columns and walls. The fixed load is used as a monotonic distributed load parallel to each beam, column, or wall and as a surface load for the deck/segment. The magnitude of the beam and column loads equals the surface area times the material weight density. The magnitude of the wall load equals wall's thickness times the wall height times the material density. The live load is calculated based on the standard loading code. The live and live loads of 20 kN/m$^2$ and 65 kN/m$^2$, respectively [36], were applied to all stories. But different loads were applied for roofs, at 54 kN/m$^2$ and 15 kN/m$^2$, respectively [26,37]. All models were fulfilled based on the rigid bed which was used for all stories. Elastic elements were considered for all beam and columns in OpenSees software which has been employed for modeling these structures. Bilin non-multi linear materials were used in order to describe the behavioral properties of elements [26]. The Krawinkler panel zone model was selected in this research (see Figure 6). The parameters of the Bilin material were obtained based on the study by Lignos and Krawinkler [38].

Panel zone distortion occurs mainly in shear because of contrary moments in the columns and beams. Panel zone modeling is explicitly performed through the method described by [39]. In this approach, modelling of the panel zone is performed in a rectangular pattern with eight very stiff elastic beam-column components. Shear deformation in the panel zone is shown by one rotational source (Figure 6).

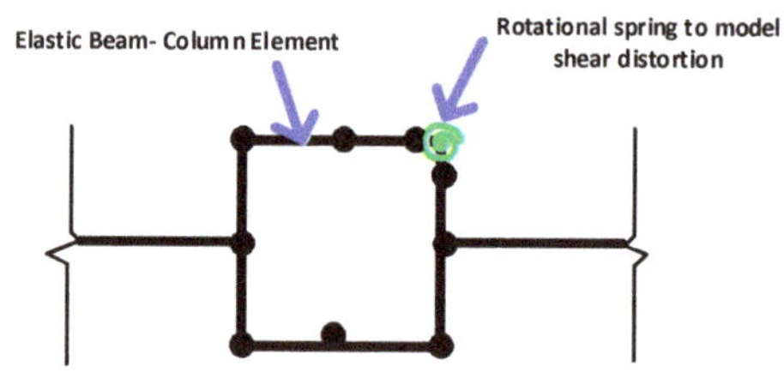


**Figure 6.** Conceptual display of an ordinary panel zone [40].

The Bilin material is defined as the corrected deterioration model of Ibarra–Medina–Krawinkler with a bilinear disturbed response. Figure 7 presents the components of the Bilin material. Lignos and Krawinkler presented the relationships between the variables [38]. The 3-, 5- and 8-story buildings have fundamental periods equal to 0.48, 0.91, and 0.78 s, respectively.

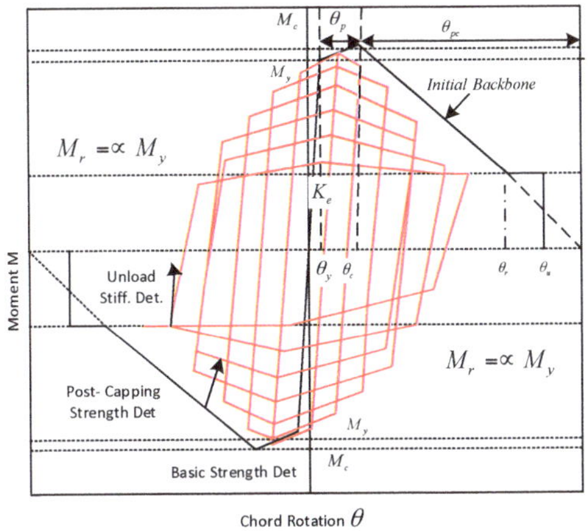


**Figure 7.** The corrected deterioration model of Ibarra–Medina–Krawinkler [38].

For simulating the nonlinear behavior of the investigated structures, their modeling was performed by using elastic beam-column components in association with rotational sources. According to the corrected IMK (Ibarra–Medina–Krawinkler) deterioration approach, the sources are supposed to represent a bilinear hysteretic reaction. Rotational springs are employed to represent the nonlinear behavior of moment frames using the concentrated plasticity concept.

A rotational spring, located at the center of the reduced beam sections (RBS) and connected to the panel zone through an elastic beam-column member, is used to model the plastic hinge. Each frame member is modeled as an elastic element with two rotational springs at either end to connect it to other elastic elements. Therefore, in order to simulate the stiffness of the actual frame member, the stiffness of different components of the structure should be adjusted [41].

### 3.2. Records Selection

Choosing the appropriate record is one of the most important parts of time history analysis. In this research, 12 earthquake records have been selected for nonlinear time history analysis, meanwhile, near-field earthquakes have the effects of fling-step earthquakes. The characteristics of these seismographs have been presented in Table 2.

## 4. Evaluating the Seismic Response of Structures

The 5% damped response spectrum should be produced and combined for each couple by applying the square root of the sum of the square (SRSS) approach, so that a specific spectrum is built for each couple. Ground movements should be measured such that the mean value of their SRSS spectra does not decrease less than 1.4 times the standard design spectra during $0.2T$–$1.5T$ seconds, where $T$ indicates [34].

The present research analyzed 36 nonlinear time histories according to 12 chosen ground motions. The maximum inter-story drift ratio was used as the first criterion for assessing the seismic demand on the structures. Figure 8 shows the maximum drift ratio stemming from the nonlinear time history analysis under the simultaneous effects of the horizontal as well as vertical components of ground motions for buildings with a special moment framework.

**Table 2.** Earthquake records selection [8].

| No | #Earthquakes | Year | Event | Mech [a] | Station | Comp | Site Class | Mw | PGA (g) | PGV (cm/s) | PGD (cm) | Fling Disp. (cm) |
|---|---|---|---|---|---|---|---|---|---|---|---|---|
| a) Far-Field Recordings Vertical | | | | | | | | | | | | |
| 1 | #EQF$_V$1 | 1992 | Big Bear | SS | Desert Hot Spr. | 90 | Soil | 6.4 | 0.11 | 10 | 1.21 | - |
| 2 | #EQF$_V$2 | 1952 | Kern County | TH/REV | Taft | 111 | Soil | 7.5 | 0.11 | 7 | 5.76 | - |
| 3 | #EQF$_V$3 | 1989 | Loma Prieta | OB | Cliff House | 90 | Stiff Soil | 7.0 | 0.06 | 0.07 | 1.43 | - |
| b) Far-Field Recordings Horizontal | | | | | | | | | | | | |
| 1 | #EQF$_H$1 | 1992 | Big Bear | SS | Desert Hot Spr. | 90 | Soil | 6.4 | 0.23 | 19.14 | 4.53 | - |
| 2 | #EQF$_H$2 | 1952 | Kern County | TH/REV | Taft | 111 | Soil | 7.5 | 0.18 | 17.50 | 8.79 | - |
| 3 | #EQF$_H$3 | 1989 | Loma Prieta | OB | Cliff House | 90 | Stiff Soil | 7.0 | 0.11 | 19.79 | 5.02 | - |
| c) Near-Field Recordings (Fling-step) Horizontal | | | | | | | | | | | | |
| 1 | #EQN$_H$1 | 1999 | Chi-Chi | TH | TCU084 | NS | Soil | 7.6 | 0.42 | 42.63 | 64.91 | 59.43 |
| 2 | #EQN$_H$2 | 1999 | Chi-Chi | TH | TCU052 | NS | Soil | 7.6 | 0.44 | 216 | 709.09 | 697.12 |
| 3 | #EQN$_H$3 | 1999 | Chi-Chi | TH | TCU0129 | NS | Soil | 7.6 | 0.61 | 54.56 | 82.70 | 67.54 |
| d) Near-Field Recordings (Fling-step) Vertical | | | | | | | | | | | | |
| 1 | #EQN$_V$1 | 1999 | Chi-Chi | TH | TCU084 | NS | Soil | 7.6 | 0.32 | 25 | 13.24 | - |
| 2 | #EQN$_V$2 | 1999 | Chi-Chi | TH | TCU052 | NS | Soil | 7.6 | 0.34 | 38.8 | 24.39 | - |
| 3 | #EQN$_V$3 | 1999 | Chi-Chi | TH | TCU0129 | NS | Soil | 7.6 | 0.19 | 24.4 | 154 | - |
| e) Near-Field Recordings (Non Fling-step) | | | | | | | | | | | | |
| 1 | #EQN1 | 1989 | Loma Prieta | OB | LGPC | 00 | Stiff Soil | 7.0 | 0.56 | 94.81 | 41.13 | - |
| 2 | #EQN2 | 1994 | Northridge | TH | Olive view | 360 | Soil | 6.7 | 0.84 | 130.37 | 31.72 | - |
| 3 | #EQN3 | 1992 | Cape Mendocino | TH | Petrolia | 90 | Stiff Soil | 7.1 | 0.66 | 90.16 | 28.89 | - |

[a] Faulting Mechanism = TH: Thrust; REV: Reverse; SS: Strike-slip; OB: Oblique.

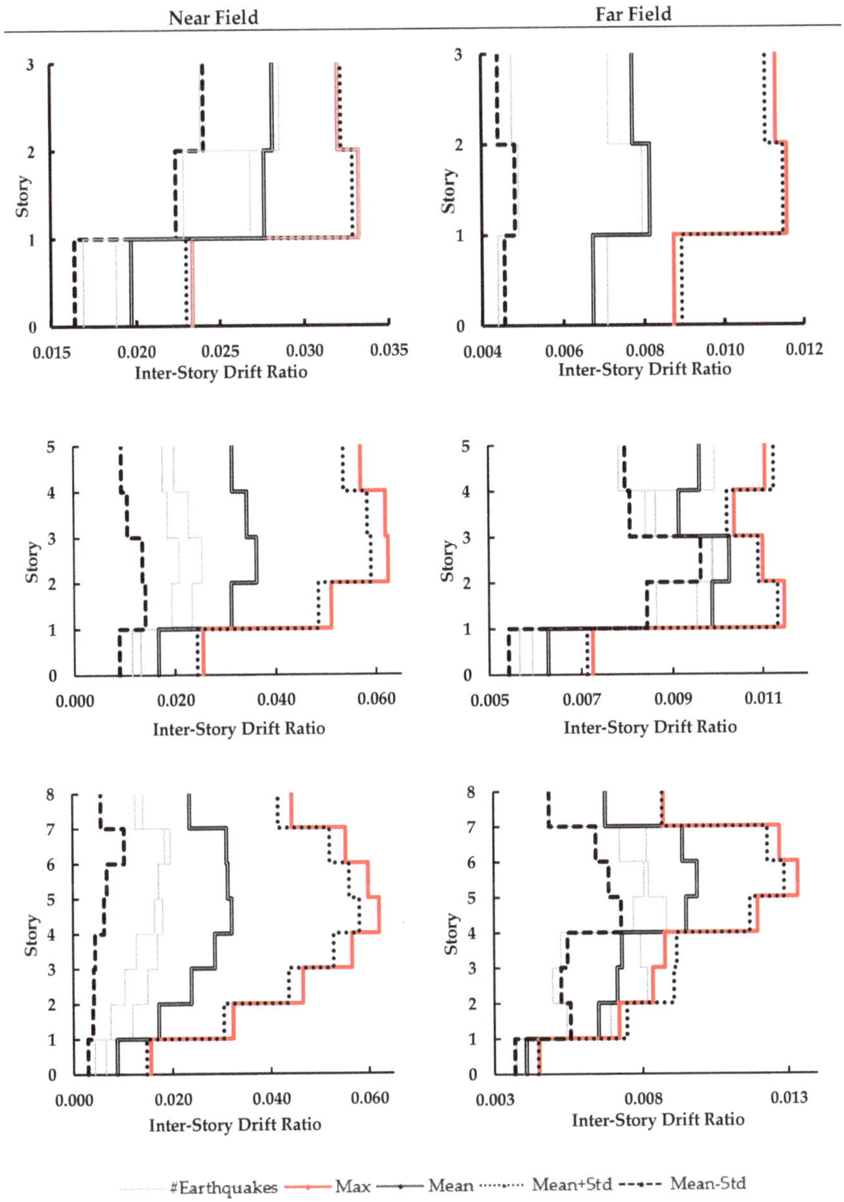


**Figure 8.** The maximum inter-story drift ratio of stories in studied frames under the simultaneous effects of the vertical and horizontal constituents of far- and near-field ground motions with fling-step records.

The graphs suggest that, in a three-story building, the maximum inter-story drift ratio was 0.03 under the simultaneous effects of the horizontal and vertical constituents of near-field earthquakes with fling-step and 0.01 for far-field earthquakes. These maximum ratios are associated with Big Bear (#(#$EQF_{(V+H)}1$) and Chi-Chi-52 records (#$EQN_{(V+H)}2$), and emerged on the second floor of the structure in both earthquakes. In addition, the maximum inter-story drift ratio was three times larger in the near-field earthquake compared to that in the far-field earthquake.

This maximum ratio was 0.01 subjected to the influence of far-field records and 0.06 subjected to the effect of near-field records in a five-story building, which were respectively associated with Chi-Chi-52 #$EQN_{(V+H)}2$) and Kern County (#$EQF_{(V+H)}2$). The inter-story drift ratio was maximized on the second floor of the structure subjected to the influence of Chi-Chi-52 (#$EQN_{(V+H)}2$) and on the third floor in Kern County (#$EQF_{(V+H)}2$). This maximum ratio was six times larger in the near-field earthquake compared to in the far-field earthquake. Although this trend is similar in both buildings, the inter-story drift ratio was maximized on the sixth floor of the five-story building in the near-field earthquake of Big Bear (#$EQF_{(V+H)}1$) and on the fifth floor of the eight-story building in Chi-Chi-52 (#$EQN_{(V+H)}2$). Table 3 compares the average maximum inter-story drift ratio in all three structures for far-field and near-field earthquakes. In this table, "Near/Far" denotes the ratio of near- and far-field.

**Table 3.** Comparison of the average of seismic response parameters of studied frames under to the simultaneous effects of the horizontal and vertical constituents of far- and near-field earthquakes with fling-step records.

| Story | Field | Inter-Story Drift Ratio $(Max)_{Ave}$ | Near/Far | $Displacement_{Ave}$ (m) | Near/Far |
|---|---|---|---|---|---|
| 3 | Near | 0.03 | 3 | 0.23 | 3.29 |
| | Far | 0.01 | | 0.07 | |
| 5 | Near | 0.04 | 4 | 0.47 | 3.36 |
| | Far | 0.01 | | 0.14 | |
| 8 | Near | 0.03 | 3 | 0.58 | 3.41 |
| | Far | 0.01 | | 0.17 | |
| | | **$Axial\ Force_{Ave}$ (kN)** | | **$Moment_{Ave}$ (kN·m)** | |
| 3 | Near | 500.8 | 2.16 | 552.7 | 2.64 |
| | Far | 232.1 | | 209.5 | |
| 5 | Near | 1173.5 | 1.75 | 1078.2 | 2.35 |
| | Far | 672.6 | | 458.3 | |
| 8 | Near | 2250.9 | 1.55 | 1376.1 | 1.9 |
| | Far | 1448 | | 723.1 | |
| | | **$Acceleration_{Ave}$ ($m/s^2$)** | | **$Velocity_{Ave}$ (m/s)** | |
| 3 | Near | 17.76 | 2.09 | 1.42 | 2.12 |
| | Far | 8.48 | | 0.67 | |
| 5 | Near | 18.57 | 1.74 | 2.06 | 1.93 |
| | Far | 10.67 | | 1.07 | |
| 8 | Near | 20.48 | 2.02 | 2.39 | 2.06 |
| | Far | 10.15 | | 1.16 | |

In the three-story building, the maximum horizontal displacement of the roof under the simultaneous influence of the horizontal and vertical constituents was 0.28 m for the near-field earthquake and 0.10 m for the far-field earthquake. This displacement was caused by the Big Bear(#$EQF_{(V+H)}1$) and the Chi-Chi-52 (#$EQN_{(V+H)}2$) records. The displacement imposed on the structure by the near-field earthquake was 2.8 times greater than that caused by the far-field earthquake.

According to the Kern County (#$EQF_{(V+H)}2$) and Chi-Chi-52(#$EQN_{(V+H)}2$) records, in the five-story building, the maximum horizontal displacement of the roof was 0.18 m under the simultaneous effects of the horizontal and vertical constituents of the near-field earthquake and 0.14 m for the far-field earthquake with the fling-step earthquake. The horizontal displacement of the roof in the five-story building subject to the near-fault ground motions was therefore 5.86 times greater than that of the far-field earthquake.

According to the Big Bear (#$EQF_{(V+H)}1$) and Chi-Chi-52 (#$EQN_{(V+H)}2$) records, the maximum horizontal displacement of the eight-story building was 1.15 m when subjected to the influence of both components of the near earthquake and 0.21 m when subjected to the influence of the far-fault ground motions. The horizontal displacement of the roof in the near-field earthquake was therefore 5.48 times higher than that of the far-field earthquake. Table 3 compares the average maximum horizontal displacement subjected to influences of far-field and near-field earthquakes.

The maximum axial force developed in the C12 column on the first floor of the three-story building was 582.10 kN when subjected to the influences of both the horizontal and vertical constituents of the near-fault record with fling-step and 289.50 kN when subjected to the influences of the far-field earthquake (Table 1). The Big Bear (#EQF$_{(V+H)}$1) and Chi-Chi-29 (#EQN$_{(V+H)}$3) records respectively imposed these forces on the columns. The peak axial force created in the column subjected to the influence of both components of the near-field earthquake was therefore 2.01 times higher than that of the far-field earthquake.

According to the Big Bear (#EQF$_{(V+H)}$1) and Chi-Chi-52 (#EQN$_{(V+H)}$2) records, the peak axial force developed by the near-fault record in the C13 column of the first floor of the eight-story building was 3109.9 kN and that by the far-field earthquake was 1598.50 kN. The maximum force developed by the near-field earthquake in the column was therefore 1.95 times higher than that developed by the far-field earthquake. Table 3 compares the average maximum axial force in the column subjected to the influence of far-field and near-field ground motions.

According to the Big Bear (#EQF$_{(V+H)}$1) and Chi-Chi-52 (#EQN$_{(V+H)}$2) records, the maximum moment developed in the C11 column of the first floor of the three-story building subjected to the simultaneous effects of the horizontal and vertical constituents of the near-fault records with a fling-step earthquake was 613.60 kN·m, and that of the far-field earthquake was 265.60 kN·m. The maximum moment developed by the near-field earthquake was therefore 2.31 times higher than that of the far-field earthquake.

According to the Kern County (#EQF$_{(V+H)}$2) and Chi-Chi-52 (#EQN$_{(V+H)}$2) records, the maximum moment developed in the C13 column subjected to the effects of the horizontal and vertical constituents of the near-field earthquake was 1578.00 kN·m and that of the far-field earthquake was 522.50 kN·m. The moment created in the column by the near-field earthquake was therefore 3.02 times higher than that created by the far-field earthquake.

According to the Big Bear (#EQF$_{(V+H)}$1) and Chi-Chi-52 (#EQN$_{(V+H)}$2) records, the maximum moment developed in the C13 column of the first floor of the eight-story building by the near-field earthquake was 2316.90 kN·m and that developed by the far-field earthquake was 796.80 kN·m. The maximum axial force developed by the near-field earthquake was therefore 2.91 times higher than that developed by the far-field earthquake. Table 3 compares the average maximum moment of the column subjected to the influence of far-field and near-field records.

According to the Big Bear (#EQF$_{(V+H)}$1) and Chi-Chi-52 (#EQN$_{(V+H)}$2) records, the maximum story shear developed in the three-story building subjected to the influences of the near-field and far-field earthquakes was 3940.80 kN and 3519.30 kN, respectively. The maximum story shear developed in the three-story building by the near-field earthquake was therefore 12% higher than that developed by the far-field earthquake.

According to the Big Bear (#EQF$_{(V+H)}$1 (and Chi-Chi-52 (#EQN$_{(V+H)}$2) records, the maximum story shear developed in the five-story building was 11,727.70 kN when subjected to the near-field earthquake and 9929.40 kN when subjected to the far-field earthquake. The maximum story shear developed by the near-field earthquake was therefore 1.18 times higher than that developed by the far-field earthquake.

According to the Big Bear (#EQF$_{(V+H)}$1) and Chi-Chi-29 (#EQF$_{(V+H)}$3) records, the maximum story shear created in the eight-story building by the near-field earthquake was 42,066.80 kN, and by the far-field earthquake 27,993.00 kN. The maximum story shear created in the eight-story building subjected to the influence of the near-field earthquake was 1.5 times higher than that of the far-field earthquake. Table 4 compares the average maximum story shear subjected to the influence of far-field and near-field earthquakes.

**Table [illegible].** Comparison of the maximum means of the inter-story drift ratio, horizontal displacement, acceleration and velocity in studied frames subjected to just the horizontal component of far- and near-field earthquakes with fling-step earthquake.

| No. Story | Max Inter-Story Drift Ratio | | | Max Displacement (m) | | | Max Acceleration ($m/s^2$) | | | Max Velocity (m/s) | | | Max Column Axial Force (kN) | | | Max Story shear (kN) | | | Max Moment Force (kN·m) | | |
|---|---|---|---|---|---|---|---|---|---|---|---|---|---|---|---|---|---|---|---|---|---|
| | FF | NF | Ratio | FF | NF | Ratio | FF | NF | Ratio | FF | NF | Ratio | FF | NF | Ratio | FF | NF | Ratio | FF | NF | Ratio |
| 3 | 0.01 | 0.03 | 3 | 0.10 | 0.28 | 2.80 | 10.86 | 28.41 | 2.62 | 0.90 | 1.56 | 1.73 | 294.43 | 580.12 | 1.97 | 2777.89 | 3347.99 | 1.21 | 265.49 | 614.71 | 2.32 |
| 5 | 0.01 | 0.06 | 6 | 0.14 | 0.82 | 5.86 | 12.65 | 19.53 | 1.54 | 1.18 | 2.51 | 2.13 | 712.57 | 147,438.80 | 2.07 | 6183.12 | 8786.29 | 1.42 | 522.48 | 1578.54 | 3.02 |
| 8 | 0.01 | 0.06 | 6 | 0.21 | 1.15 | 5.48 | 12.08 | 23.20 | 1.92 | 1.40 | 3.56 | 2.54 | 1579.56 | 299,140.4 | 1.89 | 17,242.75 | 23,608.33 | 1.37 | 798.33 | 2305.15 | 2.89 |

**Table 8.** Comparison of the maximum mean of the inter-story drift ratio, horizontal displacement, acceleration and velocity in studied frames subjected to just the horizontal (H) component of the far- and near-field earthquakes with the simultaneous effects of the vertical and horizontal (H+V) components with fling-step earthquakes.

| No. Story | Max Inter-Story Drift Ratio | | | Max Displacement (m) | | | Max Acceleration ($m/s^2$) | | | Max Velocity (m/s) | | | Max Column Axial Force (kN) | | | Max Story Shear (kN) | | | Max Moment Force (kN·m) | | |
|---|---|---|---|---|---|---|---|---|---|---|---|---|---|---|---|---|---|---|---|---|---|
| **Near Field** | | | | | | | | | | | | | | | | | | | | | |
| | H | H+V | Ratio | H | H+V | Ratio | H | H+V | Ratio | H | H+V | Ratio | H | H+V | Ratio | H | H+V | Ratio | H | H+V | Ratio |
| 3 | 0.03 | 0.03 | 1 | 0.28 | 0.28 | 1 | 28.41 | 28.38 | 1 | 1.56 | 1.57 | 1 | 580.12 | 582.11 | 1 | 3347.99 | 3940.83 | 1.18 | 614.71 | 613.56 | 1 |
| 5 | 0.06 | 0.06 | 1 | 0.82 | 0.82 | 1 | 19.53 | 19.49 | 1 | 2.51 | 2.50 | 1 | 1474.39 | 1534.49 | 1.04 | 8786.29 | 11,727.70 | 1.33 | 1578.54 | 1578.02 | 1 |
| 8 | 0.06 | 0.06 | 1 | 1.15 | 1.15 | 1 | 23.20 | 23.34 | 1 | 3.56 | 3.56 | 1 | 2991.40 | 3109.87 | 1.04 | 23,608.33 | 42,066.76 | 1.78 | 2305.15 | 2316.94 | 1 |
| **Far Field** | | | | | | | | | | | | | | | | | | | | | |
| | H | H+V | Ratio | H | H+V | Ratio | H | H+V | Ratio | H | H+V | Ratio | H | H+V | Ratio | H | H+V | Ratio | H | H+V | Ratio |
| 3 | 0.01 | 0.01 | 1 | 0.10 | 0.10 | 1 | 10.86 | 10.81 | 1 | 0.90 | 0.91 | 1.01 | 294.43 | 289.49 | 0.98 | 2777.89 | 3519.26 | 1.27 | 265.49 | 265.62 | 1 |
| 5 | 0.01 | 0.01 | 1 | 0.14 | 0.14 | 1 | 12.65 | 12.74 | 1.01 | 1.18 | 1.18 | 1 | 712.57 | 712.57 | 1 | 6183.12 | 9929.37 | 1.61 | 522.48 | 522.54 | 1 |
| 8 | 0.01 | 0.01 | 1 | 0.21 | 0.21 | 1 | 12.08 | 12.02 | 1 | 1.40 | 1.40 | 1 | 1579.56 | 1598.53 | 1.01 | 17,242.75 | 2799.27 | 1.62 | 798.33 | 796.82 | 1 |

## 5. Conclusions

The fling-step property due to a stationary ground displacement is generally performed through a combination of a unilateral velocity pulse and a uniform stage in the displacement time interval. The separate stage in the displacement time interval happens along the fault slip (i.e., in the direction of strike and dip for strike- and dip-slip occurrences, respectively).

According to existing studies, near-field earthquakes are characterized differently from far-field earthquakes in terms of the frequency content and amplification range. Thus, it is necessary to study the effects of such parameters on structures. With respect to near-fault fields, forward directivity and fling-step pulses may appear in the case of each component with a parallel or perpendicular position to the fault line, either separately or together in one record component. This study evaluates the seismic behaviors of three-floor, five-floor and eight-floor special steel bending frames that undergo near-field earthquakes with fling-step effects and also far-field earthquakes. This study also modeled the panel zone in order to render a more realistic behavior of the structure. Non-linear time-history analysis of the structures demonstrated that 'floor displacement', 'inter-story drift ratio', 'floor acceleration', 'axial and anchor force on the columns' and 'floor velocity' subjected to influence of the horizontal component of near-field earthquakes with a fling-step effect are greater than the corresponding values of these parameters in far-field earthquakes.

**Author Contributions:** S.S. and I.M. figured out the given idea. S.S. and A.K. presented the theory and conducted the computations. I.M. and J.W.H. investigated and confirmed the analytical approaches. J.W.H. and I.M. persuaded S.S. and A.K. to study both orientation and fling-step impacts and controlled the results of this research. All researchers considered the findings and provided the ultimate work.

**Funding:** Basic Science Research Program of the National Research Foundation of Korea (NRF) provided facilities for this research, for which the authors are grateful. Financial support of the Ministry of Science, ICT & Future Planning (2017R1A2B2010120) is highly appreciated.

**Conflicts of Interest:** The authors declare no conflict of interest.

## References

1. Bolt, B.A.; Abrahamson, N.A. 59 Estimation of strong seismic ground motions. *Int. Geophys.* **2003**, *81*, 983–1001. [CrossRef]
2. Ventura, C.E.; Archila, M.; Bebamzadeh, A.; Liam Finn, W.D. Large coseismic displacements and tall buildings. *Struct. Des. Tall Spec. Build.* **2011**, *20*, S85–S99. [CrossRef]
3. Grimaz, S.; Malisan, P. Near field domain effects and their consideration in the international and Italian seismic codes. *Boll. Geofis. Teor. Appl.* **2014**, *55*, 717–738. [CrossRef]
4. Avossa, A.M.; Pianese, G. Damping effects on the seismic response of base-isolated structures with LRB devices. *Ing. Sismica* **2017**, *34*, 3–29.
5. Carydis, P.; Lekkas, E. The Haiti Earthquake Mw = 7.0 of January 12 th 2010: Structural and geotechnical engineering field observations, near-field ground motion estimation and interpretation of the damage to buildings and infrastructure in the Port-au-Prince area. *Ing. Sismica* **2011**, *28*, 24–42.
6. Bhandari, M.; Bharti, S.D.; Shrimali, M.K.; Datta, T.K. Seismic Fragility Analysis of Base-Isolated Building Frames Excited by Near- and Far-Field Earthquakes. *J. Perform. Constr. Facil.* **2019**, *33*. [CrossRef]
7. Tajammolian, H.; Khoshnoudian, F.; Talaei, S.; Loghman, V. The effects of peak ground velocity of near-field ground motions on the seismic responses of base-isolated structures mounted on friction bearings. *Earthq. Struct.* **2014**, *7*, 1159–1282. [CrossRef]
8. Kalkan, E.; Kunnath, S.K. Effects of fling step and forward directivity on seismic response of buildings. *Earthq. Spectra* **2006**, *22*, 367–390. [CrossRef]
9. Mavroeidis, G.P.; Dong, G.; Papageorgiou, A.S. Near-fault ground motions, and the response of elastic and inelastic single-degree-of-freedom (SDOF) systems. *Earthq. Eng. Struct. Dyn.* **2004**, *33*, 1023–1049. [CrossRef]
10. Somerville, P.G.; Smith, N.F.; Graves, R.W.; Abrahamson, N.A. Modification of empirical strong ground motion attenuation relations to include the amplitude and duration effects of rupture directivity. *Seismol. Res. Lett.* **1997**, *68*, 199–222. [CrossRef]

11. Baker, J.W. Identification of near-fault velocity pulses and prediction of resulting response spectra. In Proceedings of the Geotechnical Earthquake Engineering and Soil Dynamics IV, Sacramento, CA, USA, 18–22 May 2008.
12. Akkar, S.; Yazgan, U.; Gülkan, P. Drift estimates in frame buildings subjected to near-fault ground motions. *J. Struct. Eng.* **2005**, *131*, 1014–1024. [CrossRef]
13. MacRae, G.A.; Morrow, D.V.; Roeder, C.W. Near-fault ground motion effects on simple structures. *J. Struct. Eng.* **2001**, *127*, 996–1004. [CrossRef]
14. Malhotra, P.K. Response of buildings to near-field pulse-like ground motions. *Earthq. Eng. Struct. Dyn.* **1999**, *28*, 1309–1326. [CrossRef]
15. Wong, K.K.F.; Yang, R. Comparing response of SDF systems to near-fault and far-fault earthquake motions in the context of spectral regions. *Earthq. Eng. Struct. Dyn.* **2001**, *30*, 1769–1789. [CrossRef]
16. Mohseni, I.; Lashkariani, H.A.; Kang, J.; Kang, T.H.K. Dynamic response evaluation of long-span reinforced arch bridges subjected to near- and far-field ground motions. *Appl. Sci.* **2018**, *8*, 1243. [CrossRef]
17. Zhao, D.; Liu, Y.; Li, H. Self-tuning fuzzy control for seismic protection of smart base-isolated buildings subjected to pulse-type near-fault earthquakes. *Appl. Sci.* **2017**, *7*, 1185. [CrossRef]
18. Gillie, J.L.; Rodriguez-Marek, A.; McDaniel, C. Strength reduction factors for near-fault forward-directivity ground motions. *Eng. Struct.* **2010**, *32*, 273–285. [CrossRef]
19. Jalali, R.S.; Trifunac, M.D. A note on strength-reduction factors for design of structures near earthquake faults. *Soil Dyn. Earthq. Eng.* **2008**, *28*, 212–222. [CrossRef]
20. Burks, L.S.; Baker, J.W. A predictive model for fling-step in near-fault ground motions based on recordings and simulations. *Soil Dyn. Earthq. Eng.* **2016**, *80*, 119–126. [CrossRef]
21. Farid Ghahari, S.; Jahankhah, H.; Ghannad, M.A. Study on elastic response of structures to near-fault ground motions through record decomposition. *Soil Dyn. Earthq. Eng.* **2010**, *30*, 536–546. [CrossRef]
22. Shahbazi, S.; Khatibinia, M.; Mansouri, I.; Hu, J.W. Seismic evaluation of special steel moment frames undergoing near-field earthquakes with forward directivity by considering soil-structure interaction effects. *Sci. Iran.* **2018**. [CrossRef]
23. Nastri, E.; D'Aniello, M.; Zimbru, M.; Streppone, S.; Landolfo, R.; Montuori, R.; Piluso, V. Seismic response of steel Moment Resisting Frames equipped with friction beam-to-column joints. *Soil Dyn. Earthq. Eng.* **2019**, *119*, 144–157. [CrossRef]
24. Dell'Aglio, G.; Montuori, R.; Nastri, E.; Piluso, V. Consideration of second-order effects on plastic design of steel moment resisting frames. *Bull. Earthq. Eng.* **2019**. [CrossRef]
25. Piluso, V.; Pisapia, A.; Castaldo, P.; Nastri, E. Probabilistic Theory of Plastic Mechanism Control for Steel Moment Resisting Frames. *Struct. Saf.* **2019**, *76*, 95–107. [CrossRef]
26. Mansouri, I.; Shahbazi, S.; Hu, J.W.; Arian Moghaddam, S. Effects of pulse-like nature of forward directivity ground motions on the seismic behavior of steel moment frames. *Earthq. Struct.* **2019**, in press.
27. Sánchez-Olivares, G.; Tomás Espín, A. Design of planar semi-rigid steel frames using genetic algorithms and Component Method. *J. Constr. Steel Research.* **2013**, *88*, 267–278. [CrossRef]
28. Macedo, L.; Silva, A.; Castro, J.M. A more rational selection of the behaviour factor for seismic design according to Eurocode 8. *Eng. Struct.* **2019**, *188*, 69–86. [CrossRef]
29. Mansouri, I.; Saffari, H. A new steel panel zone model including axial force for thin to thick column flanges. *Steel Compos. Struct.* **2014**, *16*, 417–436. [CrossRef]
30. Saffari, H.; Mansouri, I.; Bagheripour, M.H.; Dehghani, H. Elasto-plastic analysis of steel plane frames using Homotopy Perturbation Method. *J. Constr. Steel Research.* **2012**, *70*, 350–357. [CrossRef]
31. Li, S.; Xie, L.L. Progress and trend on near-field problems in civil engineering. *Acta Seismol. Sin.* **2007**, *20*, 105–114. [CrossRef]
32. Stewart, J.P.; Chiou, S.J.; Bray, J.D.; Graves, R.W.; Somerville, P.G.; Abrahamson, N.A. *Ground Motion Evaluation Procedures for Performance-Based Design*; PEER Report 2001/09; Pacific Earthquake Engineering Research Center, University of California Berkeley: Berkeley, CA, USA, 2001.
33. Kalkan, E.; Adalier, K.; Pamuk, A. Near source effects and engineering implications of recent earthquakes in Turkey. In Proceedings of the 5th international conference on case histories in geotechnical engineering, New York, NY, USA, 13–17 April 2004.
34. Standard2800. In *Iranian Code of Practice for Seismic Resistant Design of Buildings*, 4th ed.; Building and Housing Research Center: Tehran, Iran, 2014.

35. Etemadi Mashhadi, M.R. Development of Fragility Curves for Seismic Assessment of Steel Structures with Consideration of Soil-Structure Interaction. Master's Thesis, University of Birjand, Birjand, Iran, 2015.
36. Shahbazi, S.; Mansouri, I.; Hu, J.W.; Sam Daliri, N.; Karami, A. Seismic response of steel SMFs subjected to vertical components of far- and near-field earthquakes with forward directivity effects. *Adv. Civ. Eng.* **2019**, *2019*, 2647387. [CrossRef]
37. Shahbazi, S.; Mansouri, I.; Hu, J.W.; Karami, A. Effect of soil classification on seismic behavior of SMFs considering soil-structure interaction and near-field earthquakes. *Shock Vib.* **2018**, *2018*, 4193469. [CrossRef]
38. Lignos, D.G.; Krawinkler, H. *Sidesway Collapse of Deteriorating Structural Systems under Seismic Excitations*; Rep. No. 172; John A. Blume Earthquake Engineering Center, Stanford University: Stanford, CA, USA, 2009.
39. Gupta, A.; Krawinkler, H. *Seismic Demands for Performance Evaluation of Steel Moment Resisting Frame Structures*; Rep. No. 132; John A. Blume Earthquake Engineering Center, Stanford University: Stanford, CA, USA, 1999.
40. OpenSees. Open System for Earthquake Engineering Simulation. Available online: http://opensees.berkeley.edu:PEER (accessed on 16 May 2019).
41. Ibarra, L.F.; Krawinkler, H. *Global Collapse of Frame Structures under Seismic Excitations*; Rep. No. 152; John A. Blume Earthquake Engineering Center, Stanford University: Stanford, CA, USA, 2005.

applied sciences

MDPI

Article

# Modeling Palletized Products: The Case of Semi-Filled Bottles under Top-Load Conditions

**Ana Pavlovic [1], Cristiano Fragassa [1,*], Luca Vegliò [1], Felipe Vannucchi de Camargo [1,2] and Giangiacomo Minak [1]**

1 Department of Industrial Engineering, University of Bologna, viale Risorgimento 2, 40136 Bologna, Italy; ana.pavlovic@unibo.it (A.P.); luca.veglio@studio.unibo.it (L.V.); felipe.vannucchi@ufrgs.br (F.V.d.C.); giangiacomo.minak@unibo.it (G.M.)

2 Post-Graduation Program in Mining, Metallurgical and Materials Engineering, Federal University of Rio Grande do Sul, Rua Osvaldo Aranha 99, Porto Alegre 90035-190, Brazil

* Correspondence: cristiano.fragassa@unibo.it; Tel.: +39-346-323-2510

Received: 21 August 2019; Accepted: 7 October 2019; Published: 2 January 2020

**Abstract:** An investigation on numerical methods able to simplify the mechanical behavior of PET bottles, partially filled with liquid, under compression loadings is presented here. Compressive stress conditions on bottles are very common during their transportation and can be accompanied by large deformations and instabilities that can compromise the integrity of the pack, with the risk of significant damages. The present paper proposes two approaches, both based on finite elements, and with elastic-plastic material models properly defined with the scope of investigating the complex phenomena that take place during these loading conditions. Although not perfect in terms of accuracy, these numerical methods have been proven to be capable to predicti the transport-related integrity risks, showing results that agree with experimental data, especially during the initial phases of load compression.

**Keywords:** transport; palletized goods; damage; bottle; buckling; Polyethylene terephthalate (PET)

## 1. Introduction

The irreversible damage of goods during transportation, caused mainly by sloshing, vibration, and shock mechanical solicitations [1], is a common scenario nowadays in the logistics sector, resulting in significant economic losses to both the retailer and manufacturer [2]. The environmental aspect is particularly affected by the greenhouse gas (GHG) emissions to produce and ship goods that, being damaged, will not satisfy the customer and will end up in a landfill. On the other hand, the economic losses can be ultimately managed in some way. For instance, an increase in the price of goods, as estimated with proper methods, as in [3], can be added to ensure the economic sustainability of the industrial investment. On the contrary, the environmental cost inferred by deteriorated goods, as claimed in [4], is often definitive and articulated on multifaceted aspects, such as materials no longer usable, energy and water used for product processing, and fuel and carbon dioxide used during the transport phases.

The loss in functionality and usability of goods during transportation can be considered a non-negligible fraction of the whole number of products moved around the world. For instance, a study commissioned by Walmart [5] estimated the phenomenon in 1,176,260 pallets damaged each year, equivalent to 58,813 truckloads of product going into landfill that, at the financial level, equates to about $2400M and 18% earnings. Many leading companies are trying to identify and stem the causes of the deterioration of goods due to transport [6]. The evaluation of the economic repercussions of these phenomena through in-depth analyses [7] can lead to the identification of solutions to be adopted by the company to minimize losses.

Several recent studies [8,9] have been directed to optimize transport conditions, developing solutions for what is commonly referred to as 'secondary packaging', which does not take into consideration packaging materials that are not in direct contact with the product (e.g., blisters, sachets, boxes). In particular, in the case of bottles, the secondary packaging refers to the action of arranging the bottles together into bundles covered with stretch wrap films, positioned side by side and overlapped to form the pallets (Figure 1a). The pallets, which, in the case of bottles, can also contain tens of bundles arranged on several levels, are protected by the deposition of several layers of stretch film. This plastic and thin film is deposited with special implants, combining both the wrapping and pre-tensioning actions of the film itself (Figure 1b).

(a) (b) (c)

**Figure 1.** A multilevel system: (**a**) a single bottle; (**b**) bottles grouped in bundles; (**c**) pallet and palletizing plant.

In this way, a stabilization effect is created with respect to external dynamic stresses, which is added to the general protection offered by the presence of the plastic layer outside. Through a suitable choice of the type of film and the wrapping parameters it is possible to guarantee suitable levels of protection with respect to the most common stress conditions [10,11] that are generated during the transport phases. This is possible as long as the bottles retain their structural integrity. The external packaging can do very little if the stressors on the individual bottles are such that can damage them. This effect is not so rare, considering the attempt to use progressively lighter and slimmer bottles close to the limit conditions, in the face of always large and heavy pallets. The result is that bottles are subjected to various types of loads during transport and with various possibilities of damage, with one of the most dangerous being that which leads to buckling.

Plastic bottles have been widely used since 1989 as containers for water, milk, and other beverages. The body is usually made of PET (polyethylene terephthalate) from pre-printed plastics (Figure 2a) and then blown, while the bottle cap is made of PE (polyethylene). The form is not only conceived for an aesthetic purpose, but it also aims to guarantee adequate stiffness during transport (being stacked on top of each other) and specific features during use (Figure 2b).

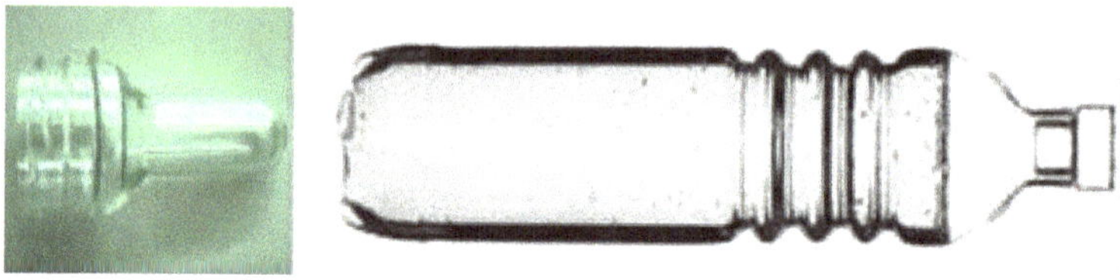

(a) (b)

**Figure 2.** (**a**) Preform in PET (polyethylene terephthalate), from which the bottle is obtained through a thermoforming process; (**b**) example of the final geometric complexity of a bottle.

The understanding of what happens during the crushing of the bottle involves a number of complications, and, to detail this complexity, it must be considered that plastic bottles are usually:

- Made of Polyethylene terephthalate (PET), a material with a restricted elastic zone: by increasing the load the plastic regime is easily reached;
- Designed in complex shapes, including reinforcing and functional elements;
- Manufactured by highly efficient and standardized processes;
- Designed with minimum thicknesses to reduce weight and cost.

The bottles are then filled, not completely, with the liquid to be contained and transported (natural water in the present case). By its turn, the partial filling represents one of the most significant reasons able to add complexity to the problem [12]. The air cushion inside the bottle creates a gas–liquid–structure interaction in response to any external mechanical (or thermal) stresses. This is the main reason why assembling a numerical model becomes challenging.

Several studies are available where experimental tests were conducted with methods and results similar to those presented in this study. For instance, in [13,14], the bottle was characterized by a monoaxial compression test, extrapolating the trend of the force as a function of the lowering. In [15], a system was specifically developed for the evaluation of the contact pressures between the bottom of the container and the fixed base, using pressure transducers in order to be experimentally correlated with a drop test. In [16], a probe for monitoring the internal pressure of the bottle was considered, stating the decrease in the volume of air during the deformation phase. In [17], a system of circumferential bands driven by a traction system was described to determine the radial resistance of the component, being able to characterize the body also orthotopically.

It must be also considered that there are widespread standard tests, for example, top-load tests (according to ASTM D 2659-95 [18], as in [17,19,20]), that lead this kind of test as a common practice, also in the case of industrial applications. The difficulty is not related to the implementation of these tests or in collecting pertinent experimental data, but in the possibility of generating a complete model (theoretical or numerical) able to interpret the information.

There are several studies that attempt to tackle this theoretical or numerical modeling through two different approaches. The first type of method, closely linked to the bottle-producing companies, consists of modeling the body according to the geometry, type, and quantity of material necessary to guarantee a predetermined resistance. In numerous studies (e.g., [21,22]) the simulations were totally focused on the minimization of the thickness to reduce the cost or, as analyzed in [23], to introduce new types of bottles to the market. Once implemented, the characteristics of the material and the discretization of the profiles with a small variable thickness represent a computational challenge, which can be addressed with the models presented in [24–26].

The second method consists of evaluating the interaction between the liquid, gas, and wall, described in [27]. This type of approach, based on both the Eulerian and Lagrangian models, is used to define the volume of liquid and air present inside the bottle, losing precision in the definition of the container itself. An example of this approach can be found in [28].

The various papers presented in [19,23] on static and dynamic simulations describe the numerical methods used for the simulation of the bottles. The effect of the interaction between water and air analyzed in [23,29] was considered only from a phenomenological point of view, that is to say, by representing it by a pressure acting on the inner surface of the container, as in [16,30,31]. In some cases, this phenomenon was modeled by means of a membrane consisting of elements that can be infinitely extended and which simulate a contact layer between the contents of the bottle and the wall. Unfortunately, such an approach, as used in [17,19], is not described in detail, making this type of simplification useless. Thus, it is possible to say that, at the moment, there are no studies that consider all the different phenomena of interest altogether. Furthermore, the complexity that has emerged would produce formulations and models that are difficult to apply in an industrial context, whereas the present interest relies on having rapid information on the risk of damage compared to the variation in shape, thickness, material, and stressors. All this seems possible only through the realization of specific experimental campaigns, one for each of the cases of interest, which represents an expensive and not always feasible approach.

## 2. Materials and Methods

### *2.1. Aim and Scope*

The present study aimed to analyze the behavior of a thin plastic bottle when subjected to vertical compression loads by integrating a numerical approach and an experimental one. For this purpose, the study presented a needful simplification of the phenomena acting on the bottle [32], characterized by a geometrically complex profile and made of PET, with the orthotropic characteristics given in Table 1. The same table also reports the main isotropic properties of PE, used for the bottle cap.

**Table 1.** Material characteristics of PET and polyethylene.

| **Polyethylene Terephthalate (PET)—Orthotropic Characteristics** | | | | | | | | | |
|---|---|---|---|---|---|---|---|---|---|
| Density | Young Modulus X dir. | Young Modulus Y dir. | Young Modulus Z dir. | Poisson's Ratio X, Y, Z | Shear Modulus XY | Shear Modulus YZ | Shear Modulus XZ | Tensile Yield Strength | Tensile Ultimate Strength |
| (kg/m$^3$) | (MPa) | (MPa) | (MPa) | (/) | (MPa) | (MPa) | (MPa) | (MPa) | (MPa) |
| 1335 | 1250 | 1345 | 1250 | 0.4 | 448 | 480 | 448 | 64 | 90 |
| **Polyethylene (PE)—Isotropic Elastic Characteristics** | | | | | | | | | |
| Density | Young Modulus | Poisson's Ratio | Tensile Yield Strength | Tensile Ultimate Strength | | | | | |
| (kg/m$^3$) | (MPa) | (/) | (MPa) | (MPa) | | | | | |
| 950 | 1100 | 0.42 | 250 | 330 | | | | | |

The bottle was considered to be partially filled with liquid, closed by the cap, constrained in the base by a friction contact, and subjected to a compression load along the longitudinal axis.

Every study took into consideration:

1. The specific tridimensional shape of the bottle (Figure 3a);
2. The variable thickness of the profile (Figure 3b);
3. The orthotropic nonlinear (elastoplastic) behavior of PET;
4. The cap rigidity, thicker than the profile, made of a more resistant PE;
5. The column of incompressible liquid, able to transmit pressure to the inner walls of the bottle;
6. The air, trapped between the free surface of the fluid and the bottle, as a compressible element.

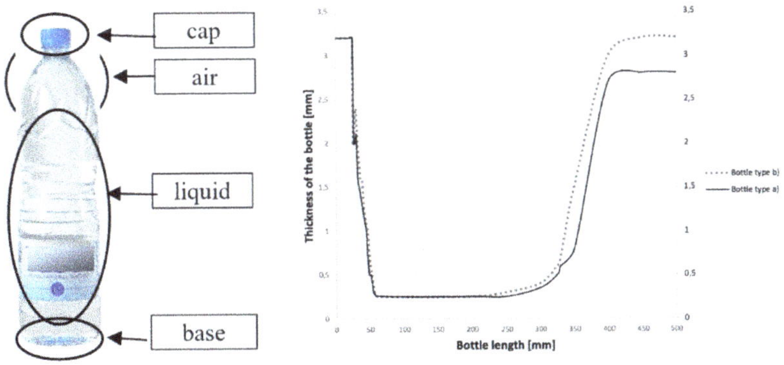


(a) (b)

**Figure 3.** Modeling the bottle: (**a**) main elements; (**b**) example of characteristic thickness [33].

With this scope, two approaches toward the simplification are proposed here.

### 2.2. Dual Approach

The first approach involved the use of an exactly faithful external geometry, as obtained from the 3D model supplied by the bottle manufacturer. In the case this geometry is not available, it could be obtained by reverse engineering, while the tendency of the thickness could be derived by analogy with respect to similar cases, available in the literature (Figure 3b). In terms of numerical discretization, the bottle was approximated by two-dimensional shell elements. The material properties were found in the vast specialized literature, taking into account the general characteristics of PET and PE. In particular, in the present case a bilinear isotropic hardening model was used for modeling PET (in accordance with [19,23,34]). The thin walls of the bottle were subjected to compression stress, to which a pressure component linked to the progressive compression of the air gap was added. The trend over time of the pressure during the test could be estimated through the equation of real gases (or, given the low pressures involved, through the perfect gas equation). However, according to the literature, which foresees the explosion of the bottle in the presence of internal overpressures around 1 MPa in the case of 1.5 L bottles, a linear increase was used up to this value (Figure 4) [23]. The liquid was considered incompressible and capable both of transferring the pressure exerted by the gas to the contact walls and of adding the effect of its own pressure due to the water column (Stevino's law).

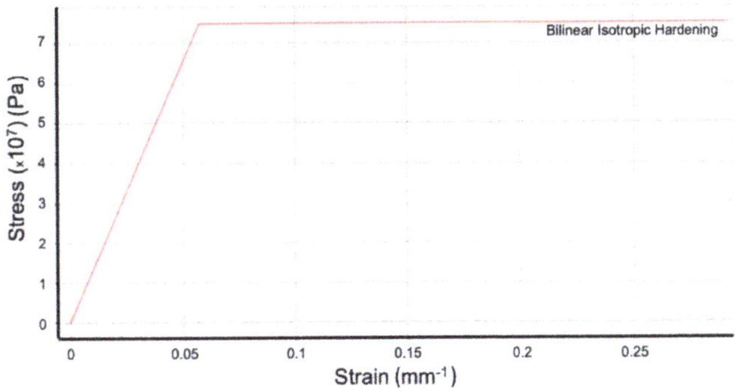


**Figure 4.** Bilinear isotropic hardening material model (in the case of 75 MPa for the yield strength and 1 MPa for the tangent modulus).

The second approach, of lower complexity, involved the use of a simplified bottle geometry, obtainable from the available databases, and chosen only by analogy with the bottles of real interest. This geometric model can be considered a single solid block, discretized by 3D solid elements (including air and liquid). The properties of the materials were assigned in such a way as to respect the macroscopic behavior of the bottle, as emerged from the experimental tests. To make this approach possible, different models and materials had to be considered and compared with the experimental evidence to verify when the top load tests were correctly reconstructed. Although this simplification may seem extreme in terms of simplification, compared to a plain investigation of the physical phenomena, sometimes it represents the only possibility. In fact, it made it possible to model the behavior of the pallet, of which the bottle represents only one of the elements. Once the model of the bottle was brought back with a black-box approach, to which numerical, literature, and experimental elements were integrated, it was possible to study the interaction between these black-boxes, grouped in bundles, which in turn interact in the composition of the pallet, all mediated through the action of polymeric coatings.

## 3. Experiments

### 3.1. Experimental Evidence

During the experimental tests, five 1.5 L water bottles (h = 310 mm, ϕ = 87 mm) were considered as specimens. Compressive tests were realized on an Instron 8032 servo-hydraulic universal testing machine with a 2517-005 load cell with ±250 kN dynamic and ±500 kN static range at a 0.25 mm/s ramp accounting for different displacements. The output data given by the testing machine were force, time, and displacement. Tests were run along the bottle axis (vertical direction), with a behaviour as shown in Figure 5. Force–displacement diagrams were also measured and displayed in Figure 6 where results from five tests are reported. These trends show an initial proportional zone (a), the bottleneck completely dented (b) and, lastly, the buckling phenomenon (c).

It can be noted that an exceptional event occurred at 8–10 mm of displacement, equal to around 2.5–3.0% of axial deformation and 0.5 kN, able to split the graph in two different regions, one with linear and the other with nonlinear behaviors. It corresponds to the indentation of the bottle, with the cap collapsing inside the neck (Figure 7a). Once the elastic response zone was passed, various phenomena became visible, increasing the load. In particular, over 2 kN, buckling happened in the specimens due to the geometry of the bottle: the tapered shape in the midways transversal plane prompted this behavior, whereas axial compression was left aside. At the end of the test, in the absence of a load, a residual deformation was also present (Figure 7b).

(a) (b) (c)

Figure 5. Predominance of buckling on single-bottle compression tests: initial part of the test with a proportional response (a), as soon as the bottleneck is completely dented (b), and near to the 3% strain region when buckling takes place (c).

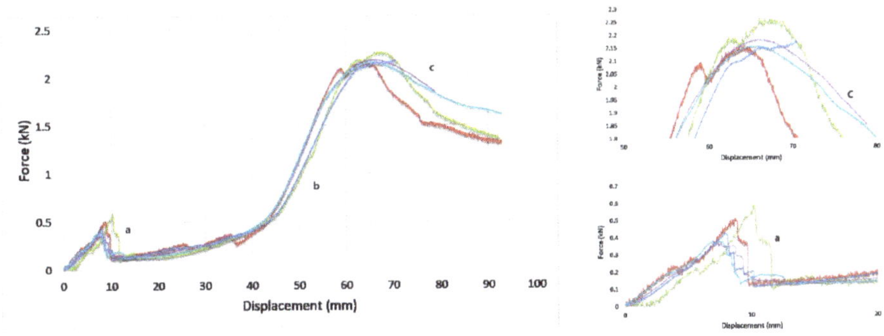


(A) (B)

**Figure 6.** Force–displacement diagrams of specimens: (**A**) whole trend and (**B**) details showing the proportional response zone (**a**), the bottleneck completely dented (**b**) and final buckling (**c**).

(**a**) (**b**)

**Figure 7.** Residual deformations with tests ending before (**a**) and after buckling (**b**).

### 3.2. An Interpretation

Within the scope to anticipate the simplification proposed later, the single bottle can be treated as a column. In this case, from the experimental measures it was possible to build an equivalent stress–strain chart to figure out an equivalent elastic modulus. The force was transformed into stress by dividing it by the circular transversal area of the bottle, with a nominal diameter of 87 mm. On the other hand, the strain was calculated based on the initial length of the specimen, i.e., the height of the bottle, equal to 310 mm.

This stress–strain diagram is reported in Figure 8a, where it is possible to verify the same unexpected variation in the trend. Before this change, a clear elastic tendency is shown until at least 1% of deformation, in which the slope can be represented using the Young's modulus (similarly to the case of the elastic response of materials). Then, the elastic modulus could be also calculated as the slope of a highly correlated linear fitting curve from 0% to 1% (Figure 8b). In other terms, if it was possible to consider the system of a bottle as an isotropic material, then the Hooke's formula should be taken into account, yielding a value equal to:

$$E = 3.08 \pm 0.47 \text{ GPa}. \quad (1)$$

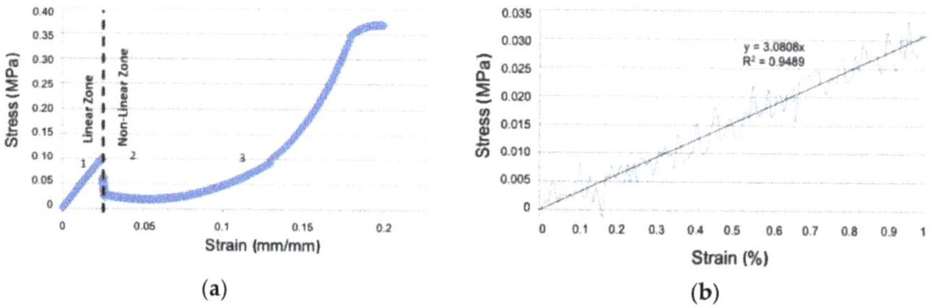


**Figure 8.** (**a**) Experimental stress–strain curve; (**b**) calculation of elastic modulus on the linear zone.

Analogously, treating the single bottle as a column, the theoretical threshold force for buckling can be calculated. This theoretical analysis, given in Equations (2)–(4), assumes the moment of inertia $I$ ($mm^4$) of a solid circular cylinder (because after the bottleneck is dented the fluid should be considered as incompressible), the elastic modulus $E$ (MPa), the column length $L$ (mm) taken as 310 mm, and finally the buckling factor $n$ which is equal to 1 in this condition where both ends are pivoted, and not fixed.

$$F = \frac{n\pi^2 EI}{L^2}, \tag{2}$$

where:

$$I = \frac{\pi d^4}{64}, \tag{3}$$

so,

$$F_{buck,theory} \cong 750\ N. \tag{4}$$

At the same time, it is evident the sharp decrease in the strain deformation diagram (shown by the transition line in Figure 8a) can be attributed to the variation in the resistant section during compression, coinciding with the collapse of the bottleneck (Figure 5b), before reaching the column of liquid. During the first compression phase, the force acts on the section corresponding to the cap, as the bottle deforms longitudinally, the resistant section widens until it reaches the maximum resistant section, which coincides with the meeting of the fluid column, equal to 5942 $mm^2$.

## 4. Numerical Models

### 4.1. Shell (2D) Model

Two-dimensional shell elements were used in this first finite elements (FE) study to evaluate the behavior of PET bottles under compression in the case of top-load tests. In particular, both a non-linear static and a buckling analysis were done by ANSYS Workbench v.18.1.

The geometry was simplified as displayed in (Figure 9a): all secondary details like fittings, chamfers, and notches were removed, since they marginally affect the system; the cap was merged to exclude contact problems; the thickness was simply defined as uniform and equal to 0.3 mm.

Four-node shell elements, 13,258 in total, were used to create the FE grid. This mesh, optimized for buckling analysis, can be observed in the area of cap and neck in Figure 9b.

In the base of the bottle all the translations and rotations were constrained (Figure 9b). The effect of water and air inside the bottle was replaced by an internal pressure, increasing over time roughly in accordance with the equation of state of perfect gases. The contact between bottle and plate was defined as a contact with friction, with a coefficient of friction equal to 0.2 representing the friction between PET and steel. Material properties were defined in line with Table 1 with a bilinear isotropic hardening model with a plastic zone starting from a yield value of 75 MPa (Figure 4).

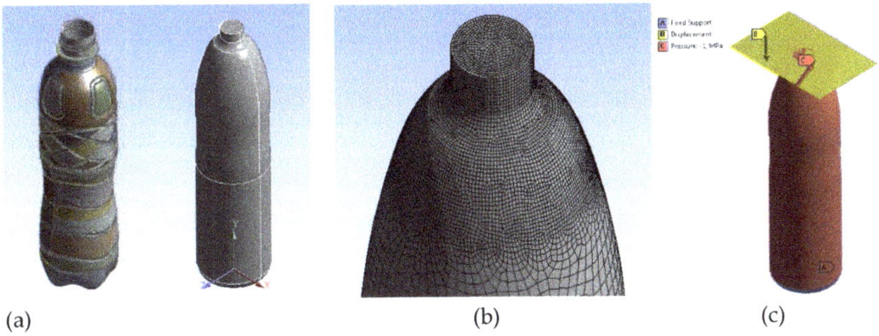


**Figure 9.** (**a**) Bottle geometry before and after simplification; (**b**) detail of geometry and mesh by shell elements; (**c**) application of forces and boundary conditions (case of non-linear static analysis).

With the scope to perform the quasi-static analysis, the bottle was loaded by a rigid plate moved by a displacement of 10 mm in the vertical (Y) direction (Figure 9c). This value was chosen by considering the maximum displacement in the elastic region as measured by top-load tests (Figure 6). In some ways, it represents the experimental evidence of the maximum load that this specific bottle can resist without moving its status to the plasticity.

The simulation of maximum load is widely used in the PET bottle industry and the vertical load test determines the weak points inside the bottles. Thus, the numerical analysis limited to this value can be extremely useful, for example, in detecting problems in the blow molding line and to check if the component failures are repeated in the same points.

Also, in this case, a non-linear structural simulation was performed with the maximum load for bottle verification in order to obtain a reliable model of an empty PET bottle. The simulation was divided into two steps: load in terms of displacement up to 10 mm, and pressure increase up to 1 MPa. Non-linearity due to geometry was taken into account through large displacements.

The stress versus strain status following a 10 mm displacement is shown in Figure 10 in terms of (equivalent) von-Mises stress (Figure 10a) and total deformation (Figure 10b). In particular, the maximum stress was 164 MPa. This stress state exceeds the yield point, showing the bottle in plastic conditions.

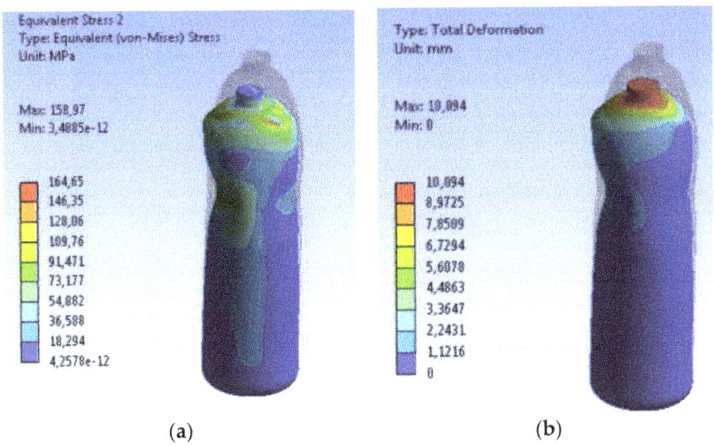


**Figure 10.** Quasi-static analysis in top-load conditions for a 10 mm vertical displacement: diagram of stress (**a**) and strain (**b**).

The buckling analysis was performed using the same conditions of simulation. It made it possible to detect the geometrical configuration of instability together with the related buckling load factor (BLF). This factor acted as a load multiplier with the scope to evaluate the buckling load for each condition of instability. When the BLF is greater than one, the structure is safe apart from when the external load is increased. It also means that when the buckling factor is less than one, the structure could collapse under a given load.

This consideration is exactly in line with what is highlighted by the results shown in Figure 11. With BLF < 1 in respect to several configurations, the thin bottle is not able to resist to a 10 mm displacement without buckling. At the same time, as reported by the simulation, the deformations seem very limited, suggesting the buckling is locally restricted.

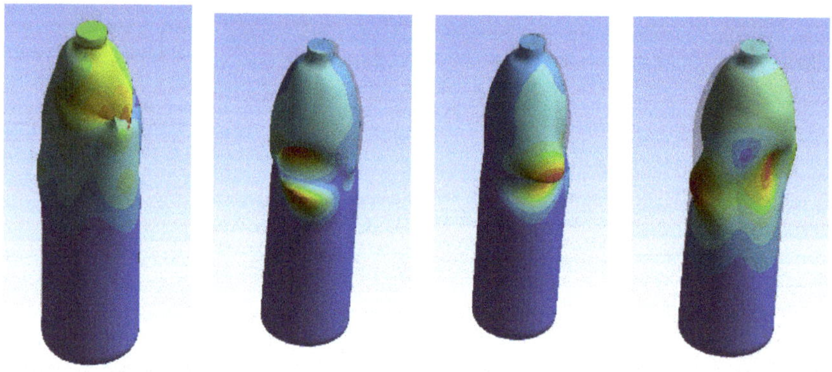

(**a**) (**b**) (**c**) (**d**)

**Figure 11.** Buckling analysis in top-load conditions for a 10 mm vertical displacement, limited to the initial four modes of instability and for a load multiplier ($k$) of: (**a**) $k = 0.19$; (**b**) $k = 0.16$; (**c**) $k = 0.12$; (**d**) $k = 0.09$.

### *4.2. Solid (3D) Model*

Three-dimensional solid elements were used to evaluate an extreme hypothesis of simplification consisting of a bottle discretized by a homogeneous object, made of a single material. The presence of liquid, air, and the plastic polymers of small thickness were not taken into account separately, but the bottle was analyzed globally, as a single solid body. This body was considered as made by an equivalent material, whose characteristics were set in order to guarantee a mechanical response in line with the experimental tests. From this perspective, the geometry should be also simplified. Starting from the 2D profile, a 3D revolution solid was generated and discretized by solid elements (Figure 12a).

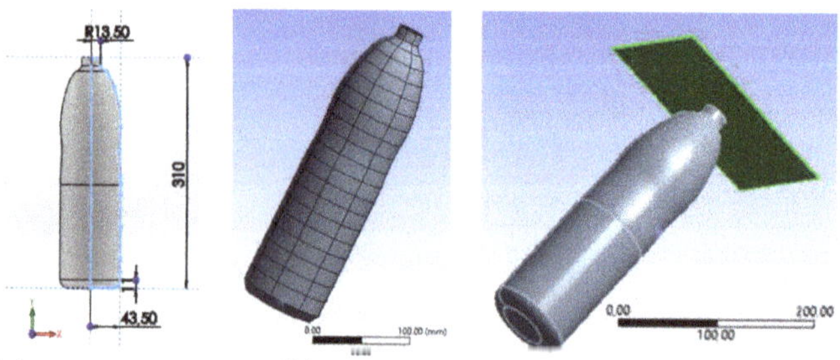


(**a**) (**b**) (**c**)

**Figure 12.** Finite Element modeling by solid elements: (**a**) 3D bottle model; (**b**) tetrahedral mesh; (**c**) boundary conditions.

A critical requirement for efficient and accurate explicit analysis is a high-quality mesh, to be also balanced with the availability of computational resources. Explicit analysis benefitted from the integration of the design phase in the ANSYS Workbench environment, which includes powerful automatic mesh generation. In general, there are two options, one is to create high-quality hex (brick) elements using multi-zone meshing, while the second one automatically uses the swept method to create hex elements. Some geometries, especially those created for manufacturing and imported from CAD data, are too complex to be swept to produce a full hex mesh. Tetrahedral elements can accurately represent small portions of a part's geometry that cannot be swept. On the contrary, tetrahedral elements can be not accurate enough in investigating large deformations and, in addition, mesh size can affect stress results. At least, different mesh density must be tuned to confirm the repeatability of this model. For this reason, it would be advisable to optimize the validity of the mesh, both in terms of element types and dimensions. At the same time, the ultimate goal of the research was to try to represent the behavior of a bottle inside a pallet in the most elementary way possible. For this reason, it was decided to limit the discretization to an automatic meshing by ANSYS tools.

In particular, quadrilateral dominant elements were used and the final mesh for explicit analysis consisted of 14,331 nodes and 13,275 elements (Figure 12a).

With respect to the first case, similarities and differences existed in the way loads and constraints were applied. The rigid plate on the top (Figure 12b) was set with a 2 mm/s speed (instead of a displacement) while bottle nodes on the bottom were fixed avoiding translations and rotations.

In an extreme simplification of this kind, which considers the entire system as homogeneous and discretized by finite elements of the same type, would be possible when a model of 'equivalent material' can offer the same responses as the real system (Figure 13).

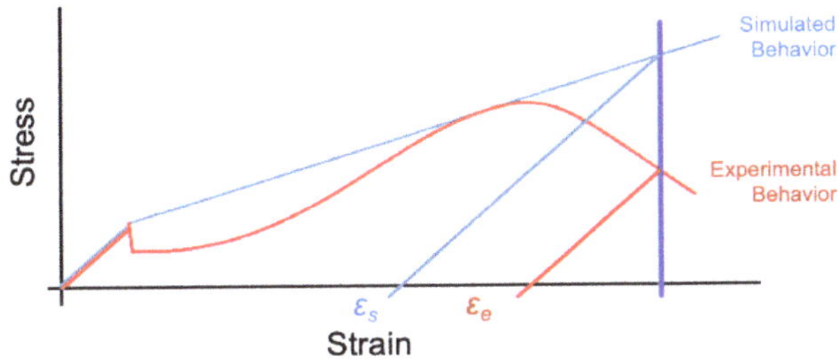


**Figure 13.** Experimental stress–strain curve discretization by linear trends, namely bilinear isotropic hardening.

Ansys Workbench 18.1 also makes it possible to insert the material behavior by additional complex methods, as in the case of the multilinear isotropic hardening or Chaboche kinematic hardening models (Figure 14). These hardening models can be generated by fitting the experimental data and are able to provide rather different responses to the simulation. In Table 2 the accuracy of predictions is evaluated by comparing real and simulated strains in the case of different hardening models.

**Table 2.** Comparing measured and simulated residual strains (ε) for different hardening models.

| Residual Strain | | Hardening Model | | |
|---|---|---|---|---|
| | **Measure ($\varepsilon_e$)** | **Bilinear Isotropic** | **Multilinear Isotropic** | **Chaboche Kinematic** |
| mm | 46.5 | 54.9 | 61.9 | 43.2 |
| Δ | | +18% | +33% | −7% |

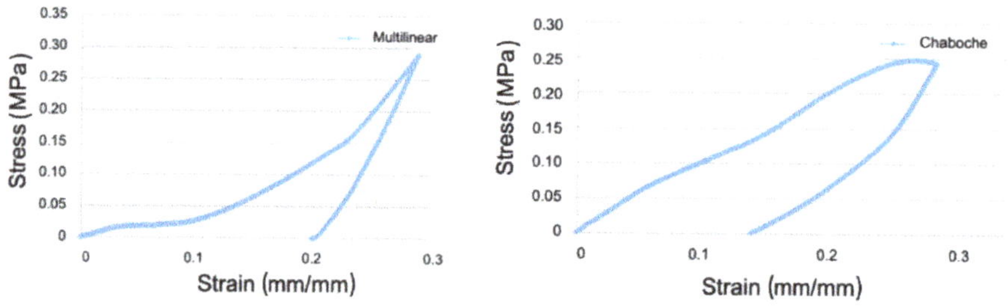


(a) (b)

**Figure 14.** Discretization by advanced hardening models: (**a**) multilinear isotropic; (**b**) Chaboche kinematic.

In particular, it is shown that simulations implementing the Chaboche kinematic hardening model are able to offer a better approximation of the system (with estimated errors in strains around 7%), as is also evident in Figure 14b, where an asymmetrical effect is correctly detected despite other models.

In Figure 15 the effect of deformation of the bottle during the top-load test with 90 mm forced displacement is shown as a simulation that adopted Chaboche kinematic hardening in comparison with the reality.

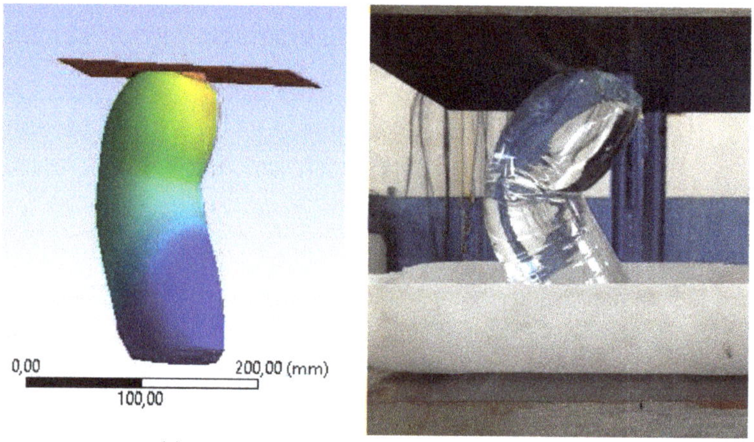


(a) (b)

**Figure 15.** Effect of a top-load test with 90 mm forced displacement: simulation (**a**) of deformations in the case of the use of a Chaboche kinematic hardening model compared with the reality (**b**).

## 5. Discussion

Disregarding the uncertainties, it is fair to assume that the experimental values of 550 N for the threshold buckling force and 3.1 MPa of elastic modulus can be used for simplified numerical simulations. Furthermore, it is also evident that both the shell model and the solid elements present interesting results, such as to make them useful for a simplified modeling of the physical system. This general aspect assumes a certain importance in consideration of the many variables involved, which make an exact solution practically unreachable. Regarding this overall complexity, it can be noted, for instance, that 188 different materials are listed in the category of unreinforced polyethylene terephthalate (PET), together with their slightly different properties to be considered.

At the same time, it should also be highlighted that the experimental and simulated values are actually closer to each other for the following reasons:

- The accurate moment of inertia is smaller because of all the tapered curves of the bottle's shape, being clearly smaller than that of whole cylinder;
- The elastic modulus used in the theoretical calculations was obtained from experiments and has a high standard deviation due to the small sampling; being potentially smaller;
- The experimental buckling force was also provided with small sampling presenting high standard deviation; being potentially higher;
- Not only the actual moment of inertia is lower, but the tapered region in the center of the bottle acts as a massive stress concentrator, also justifying low values obtained in experiments in comparison to the theoretical model.

## 6. Conclusions

The mechanical response and the stability conditions of a single bottle under compression loads were considered in this paper, as the first fundamental step of a multilevel approach aiming at the complete discretization of a pallet behavior (Figure 16).

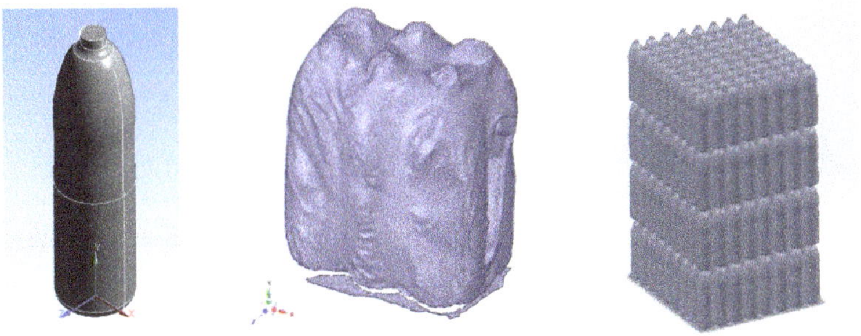

(a) (b) (c)

**Figure 16.** Discretization by a multilevel approach: (**a**) single bottle; (**b**) bottles in bundles; (**c**) pallet.

In terms of methodology, experimental measures were used to investigate two different models of discretization, each one able to take into account some essential aspects of the overall problem, simplifying the others. Their assumptions were validated through a comparison between the numerical predictions and the experimental evidence. In particular, a top load test was used on five partially filled bottles, which revealed a mechanical behavior articulated in different phases.

As general results, it is possible to say that both models acceptably predicted the initial phases of bottle deformation, characterized by an elastic response of the system. They also provided useful information regarding the effects of instability within an acceptable margin of error, making it possible to use these data for improving the bottle design.

These simplified models were also very efficient in terms of computation time, representing a first concrete step towards the development of a simplified model able to predict the static and dynamic behavior of an entire pallet during the transport phases, which represents the real challenge for the packaging industry.

**Author Contributions:** Conceptualization, C.F. and A.P.; methodology, A.P. and C.F.; software, L.V. and A.P.; validation, A.P. and C.F.; formal analysis, G.M.; investigation, F.V.d.C., L.V. and C.F.; resources, C.F.; data curation, A.P. and F.V.d.C.; writing—original draft preparation, C.F. and L.V.; writing—review and editing, C.F. and F.V.d.C.; visualization, A.P.; supervision, G.M.; project administration, G.M. All authors have read and agreed to the published version of the manuscript.

**Funding:** This research was funded by the European Union and Region Emilia Romagna inside the POR-FESR 2014–2020 action.

**Acknowledgments:** This research was implemented as part of the *'TechnoLab 4.0'* project, managed by AETNA Group S.p.a. (Italy), thanks to the financial support from the European Union and Region Emilia Romagna inside the POR-FESR 2014–2020 action.

**Conflicts of Interest:** The authors declare no conflict of interest.

## References

1. Queiroz, J.P.T.P.; Silveira, A.C.F.; Vannucchi de Camargo, F.; Vaccari, M. A brief review on palletized products transportation technology. *FME Trans.* **2018**, *46*, 101–107.
2. Vannucchi de Camargo, F.; Queiroz, J.P.T.P.; Silveira, A.C.F.; Pin, T.F.P.; Vaccari, M. Retrospect of the influence of transportation in the delivered quality of palletized product. *Int. J. Qual. Res.* **2018**, *12*, 773–788.
3. Škerlič, S.; Muha, R.; Logožar, K. A decision-model for controlling logistic cost. *Tech. Gaz.* **2016**, *23*, 145–146.
4. Colvile, R.N.; Hutchinson, E.J.; Mindell, J.S.; Warren, R.F. The transport sector as a source of air pollution. *Atmos. Environ.* **2001**, *35*, 1537–1565. [CrossRef]
5. Càvera, L. From the market laboratory, the following packaging progresses. *Leader* **2015**, *5*, 24–25.
6. Pierra, A.D.; Stewart, R.D. *International Logistics, The Management of International Trade Operations*, 4th ed.; Cicero Books Publisher: Berea, OH, USA, 2010.
7. Mussomeli, A.; Gretchen, M.; Troy, B. Joint Industry Unsaleables Report: The Real Causes and Actionable Solutions. Available online: https://www.gmaonline.org/downloads/research-and-reports/UnsaleablesFINAL091108.pdf (accessed on 14 October 2019).
8. Kandasamy, R.; Khamis, A.B. Effects of heat and mass transfer on nonlinear MHD boundary layer flow over a shrinking sheet in the presence of suction. *Appl. Math. Mech.* **2008**, *29*, 1309.
9. Pagliosa, C.; Caires, F.C.; Bataglia, D.; Pelaquim, J. Strength Analysis of Thermoplastic Containers Using Numerical Simulation. In Proceedings of the Science in the Age of Experience 2016, Boston, MA, USA, 23–27 May 2016.
10. Fragassa, C.; Macaluso, I.; Vaccari, M.; Lucisano, G. Measuring the mechanical and climatic conditions encountered by palletized products in handling and transport. *FME Trans.* **2017**, *45*, 382–393. [CrossRef]
11. Paternoster, A.; Vanlanduit, S.; Springael, J.; Braet, J. Vibration and shock analysis of specific events during truck and train transport of food products. *Food Packag. Shelf Life* **2018**, *15*, 95–104. [CrossRef]
12. Singh, S.P.; Burgess, G. Effect of internal gas pressure on the compression strength of beverage cans and plastic bottles. *J. Test. Eval.* **1993**, *21*, 129–131.
13. Survanjumrat, C.; Chaichanasiri, E. Finite element method for creep testing of high density polyethylene lubricant oil bottle. *Kasetsart J.* **2014**, *48*, 487–497.
14. Huang, H.H.; Chen, L.W.; Lu, W.H.; Lin, W.C.; Chen, Y.C. Design and simulation analysis of lightweight HDPE milk bottle. *Polym. Polym. Compos.* **2018**, *26*, 91–98. [CrossRef]
15. Karac, A.; Ivankovic, A. Investigating the behaviour of fluid-filled polyethylene containers under base drop impact: A combined experimental/numerical approach. *Int. J. Impact Eng.* **2009**, *36*, 621–631. [CrossRef]
16. Van Dijk, R.; Van Keulen, F.; Sterk, J.C. Simulation of closed thin-walled structures partially filled with fluid. *Int. J. Solids Struct.* **2000**, *37*, 6063–6083. [CrossRef]
17. Vaidya, R. Structural Analysis of Poly Ethylene Terephthalate Bottles Using the Finite Element Method. Ph.D. Thesis, Oklahoma State University, Stillwater, OK, USA, 2012.
18. ASTM, D2659-16. *Standard Test Method for Column Crush Properties of Blown Thermoplastic Containers*; ASTM Standard on Disc.: Philadelphia, PA, USA, 2005; Volume 8, Available online: https://webstore.ansi.org/standards/astm/astmd265916 (accessed on 14 October 2019).
19. Rajakutty, A. Static and Dynamic Mechanical Properties of Amorphous Recycled Poly-Ethylene Terephthalate. Ph.D. Thesis, Oklahoma State University, Stillwater, OK, USA, 2012.
20. Van Dijk, R.; Sterk, J.C.; Sgorbani, D.; Van Keulen, F. Lateral deformation of plastic bottles: Experiments, simulations and prevention. *Packag. Technol. Sci. Int. J.* **1998**, *11*, 91–117. [CrossRef]
21. Abbès, B.; Zaki, O.; Safa, L. Experimental and numerical study of the aging effects of sorption conditions on the mechanical behavior of polypropylene bottles under columnar crush conditions. *Polym. Test.* **2010**, *29*, 902–909. [CrossRef]

22. Baude, C.K. A Preliminary Analysis of PET Barrier Technologies and Mechanical Performance Related to a 3L PET Wine Bottle. Master's Thesis, Department of Packaging Science College of Applied Science and Technology Rochester Institute of Technology Rochester, New York, NY, USA, 2008.
23. Masood, S.H.; KeshavaMurthy, V. Development of collapsible PET water fountain bottles. *J. Mater. Process. Technol.* **2005**, *162*, 83–89. [CrossRef]
24. Schneider, M.H. Investigation of the stability of imperfect cylinders using structural models. *Eng. Struct.* **1996**, *18*, 792–800. [CrossRef]
25. Vaziri, A. Mechanics of highly deformed elastic shells. *Thin-Walled Struct.* **2009**, *47*, 692–700. [CrossRef]
26. Mukherjee, S. Virtual Simulation of Top Load Performance of Plastic Bottles. In Proceedings of the 68th Annual Technical Conference of the Society of Plastics Engineers 2010 (Antec), Orlanod, FL, USA, 16–20 May 2010.
27. Reed, P.E.; Breedveld, G.; Lim, B.C. Simulation of the Drop Impact for Moulded Thermoplastic Containers. *Int. J. Impact Eng.* **2000**, *24*, 133–153. [CrossRef]
28. Suvanjumrat, C.; Puttapitukporn, T.; Thusneyapan, S. Comparison of Simulation for Fluid-Structure Interaction of Water Filled Plastic Bottles during Drop Test. Master's Thesis, Kasetsart University, Bangkok, Thailand, 2007.
29. Harshath, C.; Chandan, R.; Shashikantha, N.; Venkatesha, R. Buckling Evaluation of a Plastic Bottle Design. *Int. Res. J. Eng. Technol.* **2018**, *5*, 549–553.
30. Karalekas, D.; Rapti, D.; Papakaliatakis, G.; Tsartolia, E. Numerical and experimental investigation of the deformational behaviour of plastic containers. *Packag. Technol. Sci. Int. J.* **2011**, *14*, 185–191. [CrossRef]
31. Yuan, W.; Xie, L.H.; Zhang, G.M.; Liao, D.Z.; Lu, J.D. Structural Analysis and Design Optimization of PET Beer Bottles. *Appl. Mech. Mater.* **2013**, *312*, 21–24. [CrossRef]
32. Allahkarami, M.; Ahmed, R.U.; Bandla, S. Thickness measurement methods aiding light weighting of PET bottles. In Proceedings of the 72nd Annual Technical Conference of the Society of Plastic Engineers (SPE Antec), Las Vegas, NV, USA, 28–30 April 2014.
33. Fugen, D.; Bilal, D. An energy saving approach in the manufacture of carbonated soft drink bottles. *Procedia Eng.* **2012**, *49*, 280–286.
34. Callister, W.D.; Rethwisch, D.G. *Materials Science and Engineering: An Introduction*; John Wiley & Sons: New York, NY, USA, 2007; Volume 7, pp. 665–715.

MDPI
St. Alban-Anlage 66
4052 Basel
Switzerland
Tel. +41 61 683 77 34
Fax +41 61 302 89 18
www.mdpi.com

*Applied Sciences* Editorial Office
E-mail: applsci@mdpi.com
www.mdpi.com/journal/applsci

www.ingramcontent.com/pod-product-compliance
Lightning Source LLC
Chambersburg PA
CBHW061609210926
51850CB00009B/2261
*9783036511184*